# Lean Waste Stream

## Reducing Material Use and Garbage Using Lean Principles

Marc Jensen

# Lean Waste Stream

## Reducing Material Use and Garbage Using Lean Principles

**CRC Press**
Taylor & Francis Group
Boca Raton London New York

CRC Press is an imprint of the
Taylor & Francis Group, an **informa** business

A PRODUCTIVITY PRESS BOOK

CRC Press
Taylor & Francis Group
6000 Broken Sound Parkway NW, Suite 300
Boca Raton, FL 33487-2742

© 2015 by Taylor & Francis Group, LLC
CRC Press is an imprint of Taylor & Francis Group, an Informa business

No claim to original U.S. Government works

Printed on acid-free paper
Version Date: 20140716

International Standard Book Number-13: 978-1-4822-5317-7 (Paperback)

---

### Library of Congress Cataloging-in-Publication Data

---

Jensen, Marc.
　　Lean waste stream : reducing material use and garbage using lean principles / Marc Jensen.
　　　　pages cm
　　Includes bibliographical references and index.
　　ISBN 978-1-4822-5317-7 (paperback)
　　1. Refuse and refuse disposal. 2. Waste minimization. 3. Recycling (Waste, etc.) I. Title.

TD793.9.J46 2014
363.72'850684--dc23                                                                    2014027991

---

**Visit the Taylor & Francis Web site at**
**http://www.taylorandfrancis.com**

**and the CRC Press Web site at**
**http://www.crcpress.com**

# Contents

# Introduction

This is a book about using Lean principles to drive garbage reduction, creating more resilient, sustainable operations by aligning the economic and environmental benefits of improvements. For Lean professionals, the intention of this book is to provide a new application of the Lean toolset focused specifically on garbage reduction approaches. For those in the environmental field, the parallel intention is to clearly tie garbage reductions to cost reductions in operations. The contents of the garbage can provide indicators that point back up the supply stream to identify problems at their sources and eliminate design issues. Garbage itself is not a problem; it is only a symptom of inefficient material use or processing.

The process followed in this book outlines how to collect clear data on garbage production, analyze the contents in terms of materials and sources, and then create value stream maps and process improvement projects focused specifically on material flow for the target processes. Careful management, reduction, and reuse are emphasized, along with a minimization of material handling. Conventional recycling is only presented as a last resort to avoid landfilling.

Many specific problems will be evaluated within the realm of material flow. In some cases, new uses for waste products can be found by using the waste product as a raw material elsewhere. In other cases, garbage can be reduced by working with suppliers to better align products in the supply chain. The ultimate goal is to produce a system that handles only necessary materials, recaptures its waste products as raw materials, and produces less waste material overall. We will literally reimagine our processes by beginning with the end in mind, changing the design of our systems to avoid producing future garbage.

# About the Author

**Marc Jensen** is director of the University of Oklahoma (OU) Lean Institute in the College of Continuing Education where he teaches, coaches, develops materials, and administers programs. He specializes in applying Lean techniques to environmental sustainability and conservation efforts, driving environmentally conscious economic growth. Jensen earned his Lean Six Sigma Green Belt certification in 2009. He serves broadly in an advisory capacity on conservation, sitting on several internal OU committees and as a board member of the Oklahoma Recycling Association. In addition to his work in Lean, Jensen teaches as an adjunct in the music departments for both OU and the University of Indiana.

# 1

## The Garbage Can

Sooner or later, virtually everything we make or buy ends up either consumed or in the garbage. Garbage tells the story of our material lives and behaviors, often in a candid and unflattering light. As a by-product of our behavior, the material in our garbage is a reflection of our personal behaviors and choices. Among the stories told by garbage, the volume and type of material present in it is an indicator of the health and efficiency of the organization producing it.

The purpose of this book is to reimagine the problem of garbage through the lens of Lean manufacturing: streamlining operations, improving material flow, and reducing costs. This is done by analyzing and eliminating the contents of our garbage streams while also incorporating garbage reduction into larger efforts at Lean transformation and resource conservation. This book is intended both as a reference for dealing with specific types of solid waste problems and as a handbook to offer guide points in adapting Lean tools to the garbage problem. The goal of this work is to produce actionable, measurable improvements in the cost, volume, and toxicity of our waste streams.

Understanding our relationship to garbage can help us see how human behavior really fits into the material cycles that we rely on to live.

Efficiently running processes produce lower levels of garbage because they use their resources well. Processes producing low levels of garbage will typically also have lower operating costs for the same reason. Lower waste production is only a piece of lower operating costs, but the materials present in the garbage can tell us exactly what portion of the process can benefit from attaining higher efficiency.

We can analyze the contents of a garbage can forensically and re-create in great detail the processes, behaviors, and lifestyles that produced the garbage. The physical waste products and by-products of a process can provide a window into exactly how that process is running from a perspective that may be unavailable through other observations. This is an extremely useful perspective to gain for management.

## LOOKING INSIDE THE MAGIC BOX

The garbage can is a like magic box, and is such a ubiquitous part of American culture that the vast majority of people are oblivious to its magic. Virtually any public place includes access to one of these magic boxes: a place where one can place any object, and it becomes effectively untouchable. After an object is put in the garbage, it is taboo for anyone else except designated waste handlers to see, touch, smell, or interact with it in any way ever again. This is true in private as well as public environments. Most people do not want to interact with their own garbage, and it is a measure of privacy values that it is usually considered a great breach of trust to go through items that a guest has placed in his or her host's garbage.

The taboos surrounding garbage protect the disposal of things that are considered shameful or disgusting. It hides our secrets. Although the Supreme Court has ruled that there is no legal expectation of privacy in garbage,[1] many people react strongly when their garbage is analyzed or publicized, demonstrating that they perceive this act as an invasion of a very personal space.

The magic box (Figure 1.1) usually has a lid or narrow opening that makes it impossible to see the contents inside, and it is emptied at night by a sequestered class of staff—priests collecting the offerings—so that anything placed in the garbage effectively vanishes without a trace forever. Magic!

### Garbage and Wealth

Consider how you would answer the following set of questions about your willingness to physically take money out of the garbage:

- Would you pull a clean $1 bill out of your personal trash can if nobody else was around (i.e., a $1 bill that was neither obviously contaminated nor touching something filthy)?
- Would you pull a visibly dirty $1 bill out of your personal trash can if nobody else was around?

**FIGURE 1.1**
Public trash can with lid and small, dark opening.

- Would you pull a clean $1 bill out of a public trash can if nobody else was around?
- Would you pull a clean $1 bill out of a public trash can if a stranger could see you do it?
- Would you pull a visibly dirty $1 bill out of a public trash can if nobody else was around?
- Would you pull a visibly dirty $1 bill out of a public trash can if a stranger could see you do it?

If you answered no to all of these questions, imagine that the situation were the same, but the dollar value was $100 instead of $1.

There are two independent pressures at play here for most people: the social pressure of public degradation from digging in the trash and the deeply ingrained personal aversion to contamination from filth. As the dollar value of the currency in the garbage increases, eventually most people will hit a point where its value overcomes their aversion to filth, shame, and public humiliation. This shows that even though we value the idea of personal privacy in the garbage, our taboos around garbage are more centered on a sense of risk and reward than on moral principles. After all, digging through the trash is something that is simply below the dignity of most people, and is seen as indicative of extreme poverty.

Seeing the value of physical currency in the garbage is simple, because money is exchangeable value in its purest form. The $1 bill is not an object we prize in itself, but only as it can be exchanged for goods or services. Seeing the value of commoditized materials in the garbage is also easy—such as scrap copper or aluminum cans—although more work and commitment is required to recover the value from these materials by collecting and selling them.

The opening thought experiment shifts slightly if we ask the same questions about pulling useful objects out of the garbage. Instead of a $1 bill, imagine finding a roll of tape, box of paper clips, piece of jewelry, or any other item that you would otherwise buy and use yourself. This presents a different relationship with garbage than simply finding something with value to sell, and more people will have a harder time overcoming the garbage taboos in cases where they will physically be consuming the materials themselves. This is perhaps experienced most intensely in dealing with waste food, where the waste material is actually collected to be *eaten*.

Even in this scenario, the concrete value of the wasted materials in the garbage is still clear, so most people and businesses can make the direct connection to the dollar value of an object that has been put in the garbage. Even those who are unwilling to reclaim waste materials from the trash themselves can appreciate their wasted value as something that they would benefit from correcting.

## GARBAGE AS EMBODIED PROCESS COSTS

Waste materials in our garbage also embody time and effort. This can be more difficult to conceptualize than the direct value of materials, especially since materials that are of little inherent value can represent significant costs. For example, a stack of unread paper reports that have been created, printed, filed, and then disposed of may not have any significant material value as paper. However, the unread paper reports represent significant sunk costs in terms of labor, materials, transportation, energy, and storage space, and the additional costs incurred to dispose of or recycle them. From a Lean perspective, this is the point at which the materials in the garbage start to serve as indicators of problems in the actual workflow of processes.

This example approaches the problem of the true cost of garbage, which is only effectively dealt with by addressing its root cause rather than its symptoms. Eliminating wasteful behaviors upstream in the process will prevent the garbage from ever being generated in the first place, thus reducing costs throughout the process.

## GARBAGE VS. MATERIAL WASTE

The umbrella term *garbage* as it applies to solid waste is usually defined simply as anything that is considered worthless. However, this definition needs further clarification because "worthlessness" can have a number of causes depending on the situation. We will split this waste into two major categories to show where different behaviors drive different types of waste generation, and where different kinds of solutions may be effective. These major categories are

- **Garbage (waste from use):** Solid waste produced by the use of goods; occurring when goods are consumed, transported, used up, or worn out.
- **Material waste (source waste):** Solid waste produced in the design and manufacture of goods; occurring when goods are designed and produced.

These categories effectively split the supply chain into problems produced by the act of consumption vs. problems produced by the act of production. In virtually all of the practices that follow in this book, it is useful to divide solid waste up into these categories depending on whether the material is flowing *into* the process as raw materials, or flowing *out of* the process as waste materials. Ultimately, everyone is both a producer and consumer, and problems on one side drive problems on the other, but this is a good place from which to begin an analysis.

Real-world examples of garbage and material waste reduction are peppered throughout this book, demonstrating applications of the techniques wherever possible. The information in these examples has been drawn from projects produced through Lean training delivered by the University of Oklahoma Lean Institute, and is similar to the results that can be expected from following the waste stream reduction guidelines in this book.

## ENDNOTE

1. Local laws vary with respect to how much legal authority the police have to prohibit dumpster diving. The case establishing this precedent is *Greenwood v. California* (1988). Read more about this case at http://en.wikipedia.org/wiki/California_v._Greenwood.

# 2

## Getting Rid of Our Waste

Since the 1980s, the "reduce, reuse, recycle" mantra has been part of the American consciousness. The three Rs offer strategies for preventing a great deal of our waste from entering the landfill. Although there is a tendency to focus on the "recycle" part of the slogan, it is better to think of these three as an ordered set with recycling on the bottom: first, reduce your waste products as much as possible; then, reuse everything you can from what is left; and lastly, recycle as much of the remainder as possible. The bulk of the waste from most products' life cycles is produced before the end consumer receives them, and postconsumer recycling offers only minimal real benefits. Instead, reduction in consumption offers amplified total reductions in waste from upstream. The higher one works in the set, the more the waste is reduced.

As we explore this concept through the lens of Lean principles, our goal is to address the root causes of waste production, rather than reprocessing waste material as rework after it has been created. Although "getting rid" of waste is usually described in terms of "landfill avoidance," the intention of the Lean approach is to prevent the waste from being generated. Except for some very specific circumstances, recycling will not be the first choice in waste solutions here.

Breaking the three Rs down in more detail, this chapter examines both waste disposal and waste avoidance methods. Getting rid of waste here has two meanings, since it applies both to the disposition of existing waste (disposal) and to the prevention of new waste from occurring (avoidance). This division maps onto the idea of splitting the waste stream into garbage and material waste from the previous chapter. This division is important because there are significant behavioral differences between the two.

The methodologies discussed here are approached in order from least desirable to most desirable, with the idea being that in any situation the

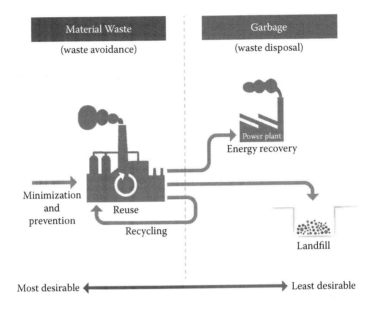

**FIGURE 2.1**

The full spectrum of methods for getting rid of waste, ranging from the most desirable (on the left) to the least desirable (on the right).

goal would be to address the waste at the highest possible level. At each successively higher level of desirability, the waste product is processed in increasingly efficient ways, until—theoretically—the need for the material producing the waste is eliminated at the final stage (Figure 2.1).

## LANDFILLING

Most waste materials are simply routed to the landfill. This is the most wasteful and often the most costly method of disposal, especially when externalized costs such as long-term environmental contamination and the impact to human health are taken into account. From a material perspective, there is no recovery of material or energy potential from landfilled materials, only the expenses associated with disposal.

The only advantage to landfilling waste material is the fact that it is simple and requires little direct labor time to process. With the exception of hazardous waste, the landfill accepts all materials, and if any special effort is required to reprocess or recover waste materials, landfilling is more expeditious. Also, in an economy with an emphasis on disposable products, landfilling is the disposition method for which many products were designed. In fact, in examining product design, it is often more costly and time-consuming to reclaim disposable materials than to landfill them. This is especially true of products that are composed of mixed material and would have to be deconstructed before recycling or other repurposing.

## ENERGY RECOVERY

In some municipalities, it is possible to recover the energy potential from waste through waste-to-energy facilities such as the Covanta facility in Tulsa, Oklahoma.[1] This approach has the advantage of providing energy for the city and also radically reducing the required landfill space, but does not make use of the materials themselves in any direct way except as fuel. Municipal waste-to-energy facilities only have a direct advantage to the waste producers where they are able to offer a lower cost for either energy or waste disposal. In doing so, energy recovery systems turn waste into a commodity, and disincentivize any efforts to reduce garbage volumes, since garbage can be perceived as a useful product (see "false commodification" below).

Please note that the energy generated by these facilities is best thought of as the recovery of a portion of the energy that was expended to create products in the first place (hence the name *energy recovery*). Many materials themselves can best be conceptualized as embodied energy. For example, wood is essentially made up of air molecules held together by the energy of stored sunlight. The wood's structure is entirely made up of embodied energy.

## RECYCLING

Recycling is the recovery of waste materials by breaking them down into their raw state and then remanufacturing them into new products (i.e., making something new out of something old). Since recycling programs typically reclaim materials that have already been refined, they have radically lower environmental footprints than sourcing the same material from virgin raw material. For example, recycling aluminum cans takes

95% less energy than extracting and refining virgin materials for the same purpose, equivalent in energy to roughly a pound of coal for every two or three cans recycled.[2] Recycling has a greater benefit than waste-to-energy systems and can be applied to more materials (greater energy efficiency vs. energy recovery).

## DOWNCYCLING AND UPCYCLING

The terms *downcycling* and *upcycling* have appeared in recent years to more precisely describe different approaches to material reclamation within the sphere of recycling.

Downcycling is the remanufacture of waste materials into a product of a lower quality.

True recycling is a closed loop, where waste products are remanufactured back into new versions of themselves. Most metal recycling has the potential to be true recycling, since the metal itself does not degrade over time with multiple smeltings. However, most fiber and plastic recycling is really downcycling because the material degrades as it is processed. For example, when recycling paper, the process of converting waste paper back into pulp tends to break the long wood fibers that are necessary to make high-quality paper. While high-quality office paper can be easily downcycled into lower-quality products such as paper towels, it can only be truly recycled into more office paper a limited number of times.

Making lower-grade paper products out of recycled paper via downcycling is invaluable. This prevents the necessity of harvesting virgin materials for these processes and also makes use of waste paper. Nevertheless, it cannot create a closed-loop system by itself.

Many recycling applications draw in a stream of recyclable materials and then use them to produce nonrecyclable outputs (such as paper towels). This is still downcycling since it adds a step of reclamation to the material's life cycle even though it ultimately leads to the landfill after that second application. Material reclamation programs that downcycle instead of recycle are useful and should not be overlooked as landfill avoidance options. Just because recycling efforts do not remanufacture materials back into the original product does not mean that they are not of value.

Upcycling straddles the line between recycling and reuse. While true recycling recovers the raw material an object is made of but destroys its form, the upcycling approach looks at the physical form of waste products with an eye toward seeing what new products they could be converted into. Upcycling is currently experiencing a trend as the chic repurposing of old products, especially via sites like Etsy and Pinterest that specialize in handcrafted products.

Although waste pallets are problematic and hard to dispose of for many businesses, there are entire cottage industries around repurposing pallet wood. Pallets can be made into all kinds of furniture, household furnishings, flooring, and wall paneling, and used for many other purposes. The use of pallet wood in remodeling or construction projects can lend a sense of hipness to a space. However, as a cautionary note, many pallets are permeated with toxic chemicals either intentionally to keep them from decomposing or by accident through spills. Pallets should be tested for chemicals before being used in applications that will directly contact people; this includes gardening.

Many upcycling efforts focus on the repurposing of either disposables or older, worn-out materials. One organization that excels at upcycling is Terracycle, Inc. (www.terracycle.com), which is based entirely on repurposing hard-to-recycle waste into new products. Terracycle specifically focuses on creating new products out of food wrappers. Companies interested in incorporating unusual recycled materials should contact Terracycle to explore collaboration with them for collection.

Recycling is often thought of only in terms of closed loops, where a material stream (such as aluminum) is endlessly remanufactured back into the same product by the same company. Both downcycling and upcycling are vital considerations that offer possibilities of reenvisioning recycling programs as *networks* of material flow connecting different kinds of organizations together. This is similar to the model of biological ecosystems where organisms evolve to occupy specific resource niches (Figure 2.2).

## RECYCLING AS A LAST RESORT

It is almost heretical to say negative things about recycling, but it is hard to ignore the fact that recycling is not the preferable way to deal with most waste materials. Recycling is only midway up the hierarchy of methods to get rid of waste, and is on the bottom of the "reduce, reuse, recycle" mantra.

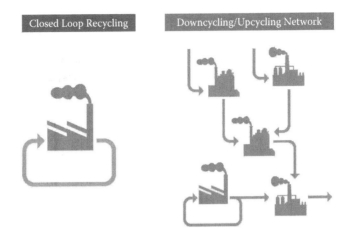

**FIGURE 2.2**
A closed loop recycling stream (left) with a single product remanufactured back into itself versus a network of interrelated materials (right) that feed into each other.

While the availability of recycling is a vital component of larger landfill diversion efforts, there are a number of major problems with it to keep in mind:

- **Internal labor costs:** Recycling is labor-intensive within the facility where material is generated. One of the major problems with recycling efforts is that they divert labor away from productive work. All of the time that a worker spends breaking down cardboard for recycling is time that he or she is not able to spend on the job. For many businesses, this problem is a barrier to recycling, and causes the business managers to accurately see that it is not in their self-interest to recycle because the costs outweigh the gains. In analyzing all processes that produce garbage, a balance has been struck that values labor more than material goods, even if that balance does not always fully account for all costs.
- **Energy:** Even though the energy requirements for recycling are typically much lower than those found in making an equivalent product out of virgin materials, the recycling process for most materials is still highly energy-intensive. Recycling requires enough energy to obliterate a waste product's form and reduce it to raw material for remanufacture, and is better avoided through the use of reusable products than through recycling disposable ones.

- **Transportation:** The energy, labor, and costly infrastructure required to transport waste materials for recycling are not necessary if the waste is never generated in the first place.
- **False commodification:** Perhaps the most insidious problem with recycling is the fact that it creates a sense of false commodification about waste. Reuse programs and material reduction efforts produce much greater benefits than recycling, but they lack recycling's high visibility through visible volumes of diverted trash.

  Recycling is a behavior that is continually reinforced as desirable and enlightened in American culture, in such a way that having a full recycling bin tends to give people a sense of good environmental stewardship and fulfilled civic duty. A higher level of thinking is required to see past this to the perspective that not generating the waste in the first place and having an *empty* recycling bin actually represents *better* environmental stewardship.

  When recyclable waste is generated, the common perspective on recycling can lead to the automatic reaction that the status quo of waste is acceptable. After all, the material is going to be recycled (e.g., "It's okay that I accidently printed 100 pages of unnecessary reports, because I'm going to recycle it").
- **Inappropriate measurement perspective:** Most major organizations that run recycling programs demonstrate their environmental stewardship by publicizing the amount of material they recycle as green public relations. Higher volumes are taken to represent better stewardship. However, these numbers almost never provide the right information needed to evaluate the success of recycling efforts.

  A recycling program is not best measured in terms of sheer tonnage, but rather in terms of what *percentage* of the available waste that tonnage represents. The capture rate of recyclable materials is the most important tool to evaluate a recycling program's effectiveness. A low volume that still represents a high percentage of the recyclable waste is a very robust program. Because the materials in the trash are usually not measured, this percentage is not available in most organizations.

  This is not a problem with recycling itself as a practice, but rather with the way it is perceived in relation to other reduction efforts (Figure 2.3).

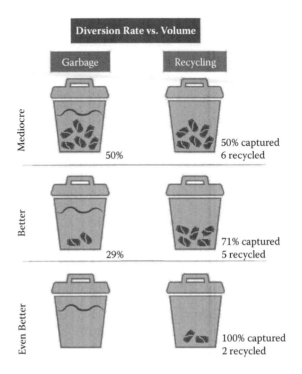

**FIGURE 2.3**

An illustration of increasingly better material disposal practices that involve progressively lower recycling rates. Because material reductions are preferable to recycling in many ways, the lower recycling rate within the context of the whole waste stream is actually a sign of better environmental practices rather than worse ones. The example on the bottom produces the least amount of recyclable material and captures it all.

- **Behavioral reinforcement:** Recycling perpetuates a disposable culture by making it acceptable to use disposable containers. Because it is easy to both recycle disposable containers and purchase products shipped in recycled disposable containers, recycling tends to kill any sense of urgency around the need to eliminate the culture of disposables in favor of reusable container systems. This creates a sense of acceptability around the disposable status quo.

  The presence of recycling does not automatically make a practice environmentally or economically friendly.

Despite recycling's problems, it is still an essential part of waste reduction efforts, and allows for the recapture of materials that would otherwise be landfilled, especially in recapturing products and materials that have truly worn out at the end of their life cycles. While it is not usually the best choice for ways to dispose of waste materials, sometimes it is at least a *better* choice than the present practice. In many situations, recycling may appear to be the best available choice for a first improvement, at least because it is simpler to implement than deeper material reduction efforts. However, the most important aspect to remember about recycling is that it is not a *sufficient* practice for true waste reduction since it only rehandles waste products that often should never have been created in the first place. For most solid waste, recycling programs should be thought of as stopgap measures that minimize landfill impact while other reduction efforts continue.

## WASTE AVOIDANCE STRATEGIES

The previous categories are based on the idea that materials in the waste stream are primarily disposable, and that the most important consideration in waste management is how best to dispose of them once they have been generated. The next three categories focus on design and sourcing, with an emphasis on not generating the waste that would then require disposal.

## REUSE

Material reuse takes one of two major forms. The first form deals with repurposing a waste product for another use in some other process. This falls on the garbage side of the waste division since it deals with waste that has already been produced (upcycling activities fall within this sphere). Garbage reuse activities are possible because different processes value *different attributes* of raw materials. Material that is no longer able to serve its original purpose may still be able to serve a second, different purpose.

One great example of this is the reuse of sand from sandblasting operations by construction companies. The sandblasting process cracks the grains of sand, which are no longer useful for sandblasting once they are pulverized to a certain size. However, this spent sand is still perfectly useful for other applications in construction where grain size is not a factor, such as the fill layer under building foundations.

The second form of reuse looks at utilizing durable and reusable products that can be recharged, refilled, repaired, cleaned, or maintained in some way to extend their life beyond their initial usage. This falls on the material waste side of the waste division, since it designs waste production out of systems rather than processing waste that has been generated. In general, reusable products reduce the need to repurpose waste materials, since they reduce the volume of waste being generated.

Durable goods are usually defined as any nonconsumable product that is designed to last for at least three years.[3] This is as opposed to consumable or disposable products that have to be frequently repurchased. Many highly durable goods have the capacity to remain functional much longer than three years and will become outmoded or obsolete before they wear out. With proper maintenance, many goods last well beyond their designed life span.

There are many single-use disposable products that masquerade as durable goods because they have a long single life (disposable ballpoint pens are an example of this). Because they last so long, it is easy to forget that they are disposable products. The key factor in determining reusability/durability is in examining the end of the product's life cycle, and whether it is designed to be disposed of or reused at that point. Even in this case, though, many disposable goods can be reused rather than discarded at the end of their natural life cycle. Do not assume that something sold as single-use disposable necessarily has to be discarded after a single use.

## MINIMIZATION AND PREVENTION

Waste minimization is typically the area where Lean is at its most effective, because it best deals with the incremental reduction of waste as it is found in existing processes. Waste minimization efforts do not typically involve the reengineering of basic process elements, but look more at the flow or layout of material, adjusting existing systems to be less wasteful.

In contrast, waste prevention efforts often reengineer processes by redesigning them so the elements that would have produced waste are no longer present. Waste prevention often involves some basic change in the nature of the process.

A good example of this distinction can be found in the efforts of supermarkets to phase out disposable shopping bags. The problem of shopping bags can be addressed in a number of different ways depending on how it is approached. Each of these is essentially a different answer to the question: How can the use of disposable shopping bags be avoided?

- Some stores make efforts to minimize waste from plastic bags by encouraging customers to use reusable grocery bags instead. The bagging process is essentially the same, but the material has changed (this topic will be revisited in the discussion about problems with reusable containers). Plastic bags are still available for those customers who fail to bring reusable ones, but their use is minimized by partially switching to the reusable product.
- Wasted bags could be minimized by training checkers in how to efficiently bag goods, so that they use only the minimal number of bags needed for each order.
- Other stores (particularly bulk outlets like Sam's Club) go beyond minimization and prevent the use of plastic bags by eliminating the bagging process. If materials are no longer bagged, the use of plastic bags disappears. To make this possible, many of the items Sam's Club sells are packed in bulk and inappropriate for plastic bags anyway, so the design of their other packaging renders bagging unnecessary.

Waste minimization and prevention are often extensions of material reuse programs, in that reuse programs tend to both minimize and prevent waste.

## ENDNOTES

1. For more information on the Covanta energy-from-waste facility, please visit its website at http://www.covantaenergy.com/facilities/facility-by-location/tulsa.aspx.
2. Laura Pritchett, "It Keeps the Heart Healthy," in *Going Green* (Norman, OK: University of Oklahoma Press, 2009), p. 110.
3. See http://en.wikipedia.org/wiki/Durable_good.

# 3

## Garbage Auditing

The first step to reducing material waste is usually to conduct a facility garbage audit. This clarifies and quantifies the content, volume, and location of materials in the waste stream. The garbage audit is a kind of Lean *gemba* walk, dissecting the waste stream through direct observation. The waste stream is first broken down into its component materials, and then the sources of those materials are analyzed to assess the best waste practices.

## PLANNING THE AUDIT

If at all possible, a garbage audit should be conducted unannounced, or at least publicized as little as possible in advance. If the event is publicized in advance, people will alter their waste behaviors—even if only unconsciously—and the resulting waste will not reflect an accurate picture of the facility's real production. Lean practitioners refer to this as the Hawthorne effect,[1] where temporary, unintentional behavioral improvements occur only because workers know that they are being observed.

## SAFETY FOR THE AUDIT

Safety is important when conducting a garbage audit. Garbage auditors may be exposed to materials that are sharp, noxious, toxic, or simply disgusting. Recommended personal protective equipment (PPE) for a trash audit includes

- **Close-toed shoes with long pants:** No flip-flops, sandals, or shorts.
- **Gloves:** Latex or nitrile gloves as an inner layer, with leather gloves as an outer layer. This will protect the hands from dangerous liquids, and also from cuts from unseen sharp objects. Needles and razor blades may turn up unexpectedly in any garbage. A thin nitrile glove does not protect against punctures. A heavy, rubberized glove may be adequate.
- **Aprons:** These are optional to protect clothing when audit participants do not wear grubby clothing.
- **Eye protection:** Goggles if necessary.
- **Mask:** Depending on the waste material, a dust mask or half-mask respirator may be appropriate.

Basic safety considerations should take into account the likely contents of the waste stream, but auditors should also be prepared for reasonable surprises. If any noxious chemicals are suspected in the waste stream, the audit should be conducted outside or in a space with good ventilation, and should always be conducted in a location where any waste liquids can drain off harmlessly.

Some people tend to be squeamish about handling garbage, even when wearing PPE, and will try avoid participating. Part of the garbage audit experience involves getting over that squeamishness and personally handling waste materials, which gives most people a new perspective on their relationship with trash. However, be sensitive to the fact that the smell of garbage will occasionally make hypersensitive people vomit (or at least *believe* that they will vomit). It is better for the whole group experience to subtly identify and remove the potential vomiters, giving them a hands-off, record-keeping role in the process so that they can still be involved.

## DUMPSTER DIVING SAFETY

In some cases, trash auditing will involve actual dumpster diving, especially if long-term waste monitoring is required. *Dumpster diving* is itself a misnomer: trash auditors should always go in feet first. Whenever possible, avoid climbing into dumpsters entirely, as there is a potential risk for injury in simply climbing back out. A long pole with a hook on the end can be used to retrieve many objects from dumpsters without bodily climbing in.

In cases where a person physically enters a dumpster, he or she should follow these basic safety guidelines:

- **Material in a dumpster may shift without warning,** injuring or pinning its occupant. Let someone know where you are going before occupying any dumpster alone, and carry a cell phone on your person while inside.
- **Be aware of the waste hauling schedule.** If possible, use your vehicle to block the dumpster you are auditing, so the dumpster cannot be dumped accidentally while you are inside.
- **Make your presence known to people who would be using the dumpster,** so that new material does not get thrown on top of you while inside. Never occupy a dumpster that is being fed through a chute from a worksite several stories overhead.
- **Wear appropriate PPE,** especially if there is any risk of exposure to materials such as fiberglass that will become airborne when moved around.
- **Be careful when removing articles from dumpsters.** Do not lean over the edge of a dumpster to pull something up and out, as this can easily lead to cracked ribs if an object is heavier than expected.

## CONDUCTING AN AUDIT AUTOPSY STYLE

Most garbage audits are conducted autopsy style as a single event, where material that has passed through the waste stream is collected and dissected and forensically analyzed by a group, as opposed to a "live" audit, where material is analyzed as it is being produced.

The following steps outline how to conduct an audit, and a worksheet is provided to simplify data collection for this process:

1. Begin planning by researching the waste hauling schedule. For example, if trash is collected on Wednesday mornings, it is most ideal to conduct the trash audit Tuesday afternoon, or even Tuesday after hours. This provides a full week's worth of trash and allows you to correlate

the trash volumes with a specific period of time in order to make better projections about average daily rates of production. This is especially true if production during the audit period can also be measured.

2. Remove all available trash from the waste stream and place it in a staging area. Unless this is done outside in a parking lot or warehouse that can be washed down afterwards, the floor of the staging area is usually covered with tarps.

3. Sort the waste into major material categories and weigh each category. The purpose of this categorization is ultimately to split materials into different streams and determine where reduction, reuse, or recycling options exist. Categories and subcategories will vary depending on the industry, but a good set of general categories to start with is

   - **Reusable materials:** Anything that could be taken directly out of the trash and put back into immediate use in an existing process in the facility.

   - **Paper:** This category includes all noncardboard recyclable paper, such as office paper, newspaper, envelopes, and possibly chipboard (noncorrugated boxes), depending on local recycling standards. Staples and paper clips are usually okay in this material, but binder clips and other larger metal attachments are not. Low-quality paper products such as paper towels, any food-contaminated paper, or any paper with a waxed coating cannot go in this category.

   - **Cardboard:** Only uncontaminated cardboard can be placed in this category. The cardboard recycling process involves solvents rather than heat, so contamination from oils and some other chemicals can render cardboard unrecyclable. For example, cardboard pizza boxes cannot be recycled due to the oil from the cheese. In fact, most recyclers will not even accept visibly clean pizza boxes due to the contamination risk.

   - **Wood:** Nails are usually okay, but larger fixtures such as hinges and handles are considered contaminants. Wood may be broken down into subcategories depending on different potential reuse or recycling options.

   - **Glass:** This is a disappearing commodity in most situations. For the purposes of sorting, include all colors of nonbroken glass bottles. Sheet glass is not accepted by most glass recyclers.

   - **Metal:** Metal items do not have to be pure, bare metal to fall in this category. Mixed metal recyclers accept relatively high levels of contaminants, although this can impact the price of the recycled goods. This is not true for recyclers of specialty metals such as copper and titanium, who demand greater purity. For example, most scrap yards will accept entire appliances—such as washing machines—as mixed scrap, without requiring that nonmetal or noniron parts first be stripped out. All of the plastic, rubber, and paint will simply burn off when the appliance is melted down. Items that are mostly metal should provisionally go in this category until exact contamination limits are clarified with the metal recycler. In general, the purer and more compact the metal, the higher the price it can be sold for per pound.

- **Recyclable plastic:** This varies based on local facilities. All recyclable plastic has a number (1–7) in the middle of the recycling triangle that refers to the chemical used to make the plastic. Most recyclers only accept plastic 1 and 2, and even then only if it is clean. Styrofoam—which is essentially puffed plastic popcorn—follows the same material numbering system. Check with the local recycler on which plastic numbers are accepted.

- **Organics:** Anything biodegradable can be placed in this category, potentially including some paper products. If organics appear in significant quantities, separate food products and more general plant products such as leaves or sawdust. This will vary widely depending on the industry.

- **Hazardous materials:** This category should be empty if material comes from the conventional waste stream, but needs to be available in case anything turns up. The term *hazardous* is defined in more detail in Chapter 12, and it can be an interesting discussion during the audit to present hazardous materials as a category and see what the participants place in it. Participants often want to place materials they feel uncomfortable with landfilling in this category, such as disposable batteries. The most likely hazardous materials will be what the Environmental Protection Agency (EPA) terms universal wastes (see Chapter 12 for more on this).

- **Garbage:** This category is the true garbage that can only go to the landfill. True garbage includes anything that does not fit into one of the other categories, as well as recyclable materials that are unrecyclable due to contamination.

These categories are only starting points that develop a very high-level view of the waste contents. Most facilities will find that they need to drill deeper into at least one of these categories in order to really paint a clear picture of waste production. For example, it may be useful to break the metal category down further into either

- Different individual metals (iron, aluminum, titanium, etc.)
- Products coming from different sources (distinctive work coming from different product lines)
- Work defects vs. true scrap
- Scrap that is from new material and scrap that is from old material being removed, for any processes that refurbish products or conduct maintenance

There is no single correct way to divide the materials. The only fitness tests the division method needs to pass are to produce a clear picture of the work that is producing the waste, and to define recoverable amounts of waste using available resources.

4. Note any interesting items, especially any durable goods or tools, as well as any items appearing in high volume. Two major issues to watch for are the shipment of air and water in the garbage, which are explored in depth in later chapters. Air simply takes up unnecessary space and reduces real container capacity, but water in containers drives up weight and has the potential to cause spills. Water also has a tendency to react with otherwise inert chemicals in waste. Containers full of water (such as drink containers full of water from melted ice in an otherwise solid waste stream) are a very different kind of problem than generating a waste stream that is primarily liquid (see Chapter 8 on dewatering waste slurry).

5. Chart out the material weights by category or specific source if possible, using the garbage audit data collection worksheet. For the most significant material components, follow up by completing the garbage interrogation worksheet in Chapter 4.

The report on the findings of the audit should also detail what percentage of the material in the garbage is recyclable.

## INCLUDE RECYCLED MATERIAL IN YOUR AUDIT

When conducting a full garbage audit, it is important to audit not only the material in the dumpster, but also whatever materials are placed in the recycling stream during the same period of time. The intention is to get a complete picture of waste generation, and in part to determine the effectiveness of recycling efforts at capturing recyclable materials.

If recycled materials are not included in the audit, the dumpster results will be ambiguous at best. For example, low levels of paper present in the dumpster could mean either that little paper is being generated or that most of it is being recycled. Without seeing both streams, it is impossible to tell.

Recycled materials are usually easier to measure because they are already segregated by type. One of the most important number sets to be produced by this audit is the comparison of weights in the garbage vs. weights in the recycling stream for each type of material recycled. For every material type recycled, chart relative weights in the recycling and garbage streams.

As an alternative to measuring recycled material directly, volumes of recycled material can sometimes be obtained from the recycling service, which may be able to report on the volumes of each material that they have received.

## VACUUM HOSES

In the special case of some manufacturing facilities, workstations may be equipped with vacuum hoses designed to suck up debris around the work surface. Where these systems exist, there can be problems with workers using the vacuum hose to dispose of scrap material rather than taking the time to dispose of it properly. This can include the disposal of raw materials such as extra screws or rivets left over from building a part as well as true production scrap. Where communal vacuum systems are used, the tracing of materials to a single workstation is virtually impossible. To be fair, where this practice takes place, it is usually done by workers who are trying to be more time-efficient, and see the recycling or restocking process as a waste of productive time.

The contents of the vacuum should be included in the analysis if applicable, possibly as a separate stream than the bulk material in the dumpster. Make sure that the contents of the vacuum being analyzed represent the same amount of time as the time window surveyed with the other waste materials for an accurate total waste picture.

Depending on the materials in the vacuum stream, a handheld magnet and sieve or screen may be useful tools to aid in sorting. PPE for this analysis should definitely include a half-mask respirator and eye protection.

## HAZARDOUS OR SENSITIVE MATERIALS

The intention of the garbage audit is to literally get to the bottom of the waste stream, which may mean exposure to some hazardous or sensitive materials depending on the industry. There should be regulations or policies governing what kinds of interactions staff are allowed to have with items that have been designated as sensitive documents, hazardous materials, or regulated medical waste. Usually in these cases, materials in these containers are locked away and may not be handled again.

If the audited waste stream includes any kind of regulated materials, treat those materials separately from the general trash audit and refer to Chapter 12 on regulated waste for more detailed instructions on how to analyze them.

If regulated materials are generated at the audited facility, be hyperaware of the potential for the regulated waste to appear in the conventional waste stream, as it may pose additional risks to the auditors. The discovery of regulated waste materials in the conventional waste stream is a significant finding of a trash audit. The cost associated with the risk of regulatory violations usually trumps the costs associated with conventional garbage disposal.

The results of this garbage analysis will provide a starting point to begin the interrogation of the waste stream for potential improvements. The worksheets in this book generally organize the data on garbage, but will almost certainly need to be tweaked based on individual situations.

## GARBAGE AUDIT DATA COLLECTION

The next few pages outline the data collection goals of the initial garbage audit, develop a high-level overview of material volumes, and measure the effectiveness of programs that handle special types of waste. A detailed case study covering a garbage audit and subsequent analysis at the University of Oklahoma has been included as Appendix A of this book.

This data collection process is also available in a more usable format as a worksheet, or may be performed based on the following tables.

## Data Collection Worksheet

Location:                                    Date:

Date range of garbage production measured:

The first three charts collect data that are merged together in the Waste Material Overview chart.

Split the conventional material stream into as many categories as reasonable. Start with major categories for the first pass, possibly drilling down later by splitting a major category into subcategories.

| Conventional Waste Stream | |
|---|---|
| Material Category | Weight Found in Garbage[a] |
| | |
| | |

[a] Include totals in the appropriate material categories under the "found in garbage" column of the Waste Material Overview chart.

| Recycling Streams | | | |
|---|---|---|---|
| Material Category | Weight of Correctly Recycled Material[a] | Weight of Any Contaminants[b] | Type of Contaminants (if any) |
| | | | |
| | | | |
| | | | |

[a] Include totals of correctly recycled material in the appropriate material categories under the "correctly recycled" column of the Waste Material Overview chart.

[b] For any improperly sorted materials found in the recycling streams, identify the material(s) and include their total weights in the appropriate material row under the "incorrectly recycled" column on the Waste Material Overview chart. For example, if paper appears in the plastic recycling stream, add the weight of that paper to the spot on the "paper" row.

| Regulated Waste Streams | | | |
|---|---|---|---|
| Material Category | Weight of Correctly Disposed Regulated Waste | Weight of Any Nonregulated Waste Found[a] | Type of Nonregulated Waste (if any) |
| | | | |
| | | | |
| | | | |

[a] For any nonregulated material improperly found in the regulated stream, identify the materials and include the totals in the appropriate material row under the "regulated waste" column on the Waste Material Overview chart below.

In the following chart, create a single row for each category of waste appearing in the other charts. For example, if *cardboard* appears in both the recycling and conventional waste, collect all of the information on cardboard into a single row.

Each material type will have a correct stream for disposal based on local availability and company policy. For each row, highlight the square in the column of correct disposition, and then calculate what percentage of the total material of that type was disposed of properly.

| Waste Material Overview | | | | | | |
|---|---|---|---|---|---|---|
| Material Category | Weight Found in Garbage | Weight Found in Recycling (correct) | Weight Found in Recycling (incorrect) | Weight Found in Regulated Waste (correct) | Weight Found in Regulated Waste (incorrect) | Percent Disposed of Correctly |
| | | | | | | |
| | | | | | | |
| | | | | | | |
| | | | | | | |

## Durable Goods, Reusable Goods, and Objects of Interest

What objects or materials caught your eye? These materials will be weighed and included in their appropriate material categories, but make note of anything interesting that showed up.

## CONDUCTING A LIVE AUDIT

For some waste streams, it is simpler to conduct a garbage audit *before* the materials reach the dumpster rather than digging them back out afterwards. This is especially useful for waste streams that produce very large volumes, very heavy materials, or very messy waste.

To conduct an audit in this manner, set up a staging area around the dumpster and have workers deposit any garbage in this staging area to be sorted and weighed before the garbage is put in the actual dumpster. The staging area can also be broken down into a number of smaller areas specifically designated for different types of waste to minimize sorting time. This method has the advantage of being able to identify the specific source of every piece of waste if desired, but has the disadvantage of potentially impacting garbage production habits by disrupting normal behaviors.

For bulky, heavy, or liquid wastes, preweighed containers can be placed in the staging area (barrels, pallets, bins, etc.). When full, these can be weighed before dumping. For more convenient wastes, hand weighing on smaller scales may be appropriate.

## ONGOING GARBAGE MONITORING

Both the autopsy and live garbage audits are single events, with the intention of getting a single snapshot of garbage over a short period of time. These cannot give information on trends or long-term performance. As part of sustaining an improvement, it is often necessary to monitor the garbage regularly over an extended period after an actual improvement has taken place. This demonstrates that the change has had the desired effect on garbage production, and in part this holds the people producing the garbage accountable by making the contents of the garbage public knowledge. Ongoing audits are sometimes part of the data collection process as preparation to make an improvement—and sometimes part of the process of sustaining improvements once they have been made.

## ENDNOTE

1. Read more on the Hawthorne effect at http://en.wikipedia.org/wiki/Hawthorne_effect.

# 4

## Interrogating the Garbage

The garbage audit provides data that allow the garbage itself to be interrogated in order to identify opportunities for improvements in the system that produces the garbage. The most basic question to ask every object in the waste stream is: Why did this material become garbage? The question that naturally follows is then: How can I prevent this material from going to the landfill? The answer to these questions usually goes back to the design process, or at the very least to purchasing.

Beginning with the highest-volume material identified in your waste audit, work through the list of questions below, and note any questions for which the answer is yes. For some materials, a yes answer may be possible to multiple questions, so continue to the end of the worksheet. A material only needs to be produced and then landfilled if the answer to *all* of these questions is no—which rarely happens.

- Is the disposal of this material out of compliance with regulations or policies in any way?
  - Garbage needs to be able to answer the most basic question of compliance. This question is first a *legal* question asking whether all controlled waste has been disposed of in compliance with regulations. It is second an *efficiency* question, asking whether the waste has been placed in an appropriate container that will neither contaminate other contents nor drive up costs.
- Could the material have been recycled/composted?
  - Very often, the majority of the waste stream is comprised of readily recyclable materials. Frequently, recycling programs already exist on site for one or more of these materials.
  - Of the materials that could not be recycled, is there a recyclable option that could replace it without adding costs?

- Is the material recyclable but contaminated in such a way that it can no longer be recycled? How did it become contaminated?
- Is the material packaging?
  - The majority of garbage is packaging, most of which is either recyclable or avoidable. One problem with packaging in the waste stream is that it tends to fill up dumpsters inefficiently. Uncollapsed boxes will very quickly fill a dumpster with air, and good containers are designed to be resistant to collapsing, requiring an investment of labor to break them down.
- Is the material expired/obsolete but otherwise nondefective?
  - If unused materials of any kind show up in the waste stream because they have expired or become obsolete, inventory and supply issues are at play. This may include problems such as departmental hoarding, or may simply represent a lack of communication among purchasers.
  - Who determines obsolescence, you or your supplier?
- Is the material a single-use disposable product that could be phased out?
  - This includes not only obvious disposable products like Styrofoam cups, but also less obvious ones like nonrefillable whiteboard markers.
- Could the material have been reused or reprocessed?
  - This will vary *widely* depending on content and imagination. The most obvious applications are things like reusing single-sided waste paper as note paper, since the material is not entirely used up at disposal. Other materials could be downcycled into new products (e.g., excess pallets into wood chips). The key issue in finding reuse options is to determine what material properties are desired in raw materials, and whether any waste materials still have those properties.
- Does the material represent process defects or errors rather than natural process by-products?
  - Scrapped raw materials or work in progress (WIP) represents a different kind of quality issue than materials that become garbage at the natural end of their life cycle. True scrap—such as pieces left over from cutting raw material down to shape—goes in this category as well. Scrap from good materials should always be considered a defect, and should be among the highest priorities for improvement efforts because it has a very clear direct cost.

- Is the material still useful as designed, but was thrown out anyway?
  - How could this still-useful item have been retained or transferred to someone else who would have been able to use it? Often materials only become garbage because they are of no further value in their current location, but they may be in high demand somewhere else.
- Is the material a durable good that was meant to last indefinitely?
  - This can be in the trash for many reasons:
    - It really has reached the end of its useful life and has worn out beyond repair. If so, why? Could this have been prevented with better maintenance?
    - It is not working now but could be repaired. What is the policy on whether something that is not working should be replaced vs. repaired?
    - It is functional but no longer looks nice, and has been replaced.
    - It is still functional but obsolete (refills, attachments, or support are no longer available for it). It is important to clarify *who* has determined that a functional product is obsolete. Forced obsolescence may simply be driven by an unscrupulous supplier who compels customers to buy a new version of its product annually even though it may not be substantially different than the previous year's.
    - It works just fine, but is no longer needed and is easier to dispose of than retain.
    - It works just fine and is in the garbage by mistake.
    - It works just fine and is being hidden in the garbage intentionally because of other problems, such as theft.

## GARBAGE INTERROGATION WORKSHEET

The following worksheet revisits the questions discussed here. The only materials that legitimately belong in the garbage are those for which the answer is no to *all* of the following questions. For any yes answers, answer follow-up questions and provide explanations. Some materials will have multiple yes answers, every applicable question should be answered rather than stopping at the first yes.

| Material: | | |
|---|---|---|
| Weight/amount uncovered in audit: | | |
| Estimated annual volume: | | |
| 1. Is the disposal out of compliance with regulations or company policies for this type of material? | Yes/No | If yes, explain why. |
| 2. Is the material recyclable or compostable? | Yes/No | If yes, go on to 2a. Also compare the amount found in the garbage with the amount recycled. |
| 2a. Is this recyclable material in the garbage because it has been contaminated? | Yes/No | If yes, explain the contaminating material. |
| 3. Is the material expired but otherwise new and nondefective? | Yes/No | If yes, what inventory controls are in place to avoid this problem? |
| 4. Is the material a single-use disposable product? | Yes/No | If yes, what durable or reusable products could replace it? |
| 4a. Is the disposable material packaging? | Yes/No | If yes, analyze the materials and different kinds of packaging involved. |
| 5. Could this material have been reused or reprocessed in your facility? | Yes/No | If yes, where and how? |
| 5a. Could this material have been reused or reprocessed by a different industry to your knowledge? | Yes/No | If yes, where and how? |
| 6. Did this material become garbage as a result of process defects rather than normal scrap and by-products? | Yes/No | If yes, what process step(s) produced this waste? |
| 7. Is this material still serviceable as designed? | Yes/No | If yes, why was it discarded? |
| 8. Is this material a durable good or tool? | Yes/No | If yes, explain why it became garbage, including the expected life span and maintenance practices for the item. |

After completing the worksheet, collect all of the explanations for the yes answers and brainstorm approaches to mitigate each of those specific problems. This list of solutions will likely address different points in the material flow, and will provide improvements to consider while mapping the process.

# 5

## *Making Improvements*

The purpose of garbage auditing is to create an accurate picture of the waste stream in order to find material efficiency problems and determine where to apply Lean tools for waste reduction, minimization, and process improvement. The garbage interrogation worksheet suggests problems to consider, but a deeper analysis of the whole process is needed in order to understand how those potential problems fit into the real process flow, and which of these problems can be translated into viable projects. Determining the best solution to implement comes only after the problem itself has been analyzed in depth and the process is well understood. Solutions will be driven by hard data. This chapter outlines some Lean tools as they are applied to this type of problem.

Many companies already have Lean programs working on transformation and continuous improvement. Where these exist, the garbage auditing process can identify garbage reduction projects that will be facilitated just like any other Lean project and woven seamlessly into larger Lean efforts. In doing this, environmental work can be understood as *efficiency* work rather than focusing solely on its ethical or political implications. Also in doing this, environmental work is given greater credibility from a business perspective since it can be seen to support core business activities, rather than being a distraction from "real" work. Both of these effects will have a tendency to reduce any opposition to environmental work, especially if that opposition is politically driven.

Likewise, many companies already have established "green teams" or other programs focused on environmental waste. These resources should be pulled into any kind of Lean environmental effort as early as possible in order to avoid duplicating existing efforts and alienating employees who are already enthusiastic supporters of sustainability work. Green teams and Lean groups are both interested in transformation and company improvements, and yet these groups rarely work together because they are often seen to be working at cross-purposes. Integrating these two teams into a single group dedicated to transformation can result in innovative solutions because each team brings a different perspective to the other group's problems.

Since the term *green* has become politicized and is polarizing to some audiences—especially in business and government—the more neutral phrase "resource conservation" has been especially successful as a replacement term that encapsulates both the environmental and economic goals of Lean. This term can be approached from both directions, with environmentally focused green teams rebranding themselves as resource conservation working groups, and business-focused Lean teams bringing in environmental considerations as an important element in overall resource management.

## IMPROVEMENT TEAMS

There are two major ways to facilitate a process improvement project: rapid improvement events (*kaizen* events) and integrated project teams. Both approaches have their advantages and are suited to different kinds of problems.

Kaizen events typically focus on very fast transformation, and are best applied when a problem revolves around redesigning flow issues. In a kaizen event, the process in question is shut down temporarily, and everyone who works on the process spends an intensive period of time (up to 1–2 weeks) redesigning and testing the process. Kaizen events often involve building mock-ups of shops to test the flow of material through them. One of the biggest advantages of the kaizen model is its flexibility. The team has the opportunity to freely rearrange process pieces and test them, along with the full support and presence of other operators. At the end of a kaizen event, everyone involved understands and buys into the new process because they designed and built it. There are usually data collection periods both before and after this intensive event in order to measure the impact of changes. One of the major disadvantages of this approach is that it involves shutting down a process while it is under reconstruction.

The Lean work model outlined in this book uses integrated project teams (IPTs). In the IPT model, a small core team facilitates the project over a period of several months. The team will work on the project part-time while also fulfilling their other job duties. The core team is usually only three to five people who represent all the major departments engaged in the project (such as purchasing, engineering, etc.). Other people who work on the process may be peripherally involved as subject matter experts (SMEs) or data collectors, but they are not at the heart of the project itself. Specific IPT roles are outlined below. Like kaizen events, IPTs have their strengths and weaknesses. The IPT model can be used to work on virtually any Lean problem. This model is best suited to projects that have more extensive data collection requirements in order to really understand the fundamental nature of the problem, and even though this model is more reliant on independent, individual work, it can often culminate in a rapid improvement event in which the team presents all of the data on the problem and involves all of the other SMEs in the process of designing the new process. Rapid improvement events in this model can include things like a short brainstorming meeting to redesign a problematic piece of paperwork together. The IPT's longer timeline can also be a weakness since it sometimes invites a loss of focus over time or allows scope creep to corrupt the project.

## PROJECT SELECTION CRITERIA

Project selection is crucial to making improvements. The results of the garbage audit will likely suggest several potential areas where Lean projects could make significant waste reductions. Determining what projects will be successful must take several important factors into account beyond the waste itself:

1. The project needs to have management support.
2. The project needs to align with the mission of the organization. The mere fact that a project reduces environmental impact or costs is not sufficient. It has to do those things in a way that ties in with the organization's mission and vision. In other words, an individual project should be seen as a tactical component of a larger strategic organizational goal. Garbage projects that tie into other efforts are much more likely to get the support they need to succeed.
3. The project needs to be focused on a clear problem.
4. The project should have clearly defined savings and return on investment as goals.
5. The project should be something that is within the control of the team, so that the members of the team have both responsibility and influence over the process.

**TABLE 5.1**

Project Selection and Prioritization Worksheet

1 = low/not good; 2 = some; 3 = OK; 4 = good; 5 = high/great.

| Criteria | Clear, Concise Problem Statement | Process Output Metrics Aligned with Top Priorities | Impacts Customer | Management Support | Process Owner Support | Data Available | Achievable in 4 Months | Minimal Capital Investment | Return on Investment | Meets Training Objectives | Project Score |
|---|---|---|---|---|---|---|---|---|---|---|---|
| Project A | | | | | | | | | | | |
| Project B | | | | | | | | | | | |
| Project C | | | | | | | | | | | |
| Project D | | | | | | | | | | | |
| Project E | | | | | | | | | | | |

6. The project should not be something that is already being worked on by another team or group from another department.
7. If a previous improvement effort has failed in this area, it is essential to understand the failure and be able to demonstrate why results are expected to be different this time.

A project that has all of these factors under control is a good candidate for a Lean improvement. If the data suggest that several problems could be candidate projects, a project selection worksheet (see Table 5.1) can be used to help narrow down the field. This worksheet can be used by either a single person or a group attempting to prioritize efforts.

## Project Selection and Prioritization Worksheet

1. List candidate problems. Review metrics, customer issues, and challenges in the business.
2. Form a clear, concise problem statement.
3. Rank each criterion for each possible project.
4. Discuss and debate the project selection score.
5. Select a project and take action to improve project scores or investigate other possible projects.

## THE PROJECT TEAM

The core project team is comprised of several important roles. Keep in mind that the ideal size of a project team is usually not more than three to five people, and because of this, a single person often covers multiple roles. Larger teams tend to lose focus and be more difficult to manage.

- **Project champion:** This person is an advocate of the project who is high enough in the management of the organization to authorize changes and allocate resources. In many larger organizations, the champion will be a VP or higher. The champion's primary role is to remove barriers, provide resources, and run interference for the team as they complete their work. The champion does not usually perform work directly on the project, but is someone to whom the project team is answerable.

- **Process owner:** This person has direct authority over the process that the team is working to improve. This person *must* be involved in the project at some level. The process owner is often a supervisor over an area.
- **Team leader:** This person directs the activities of the team, sets up meetings, manages schedules, documents the project, and interfaces with the champion. The team leader is often the process owner, or may instead be an employee who reports to the process owner. Ideally, the team leader should have experience with project management.
- **Team members:** These people carry out project assignments and help with administration and logistics as required. Team members should be selected based on their involvement with the problem, and represent different functional roles.
- **Subject matter experts (SMEs):** These are any other people who work in or on the area impacted by the project. They may not be directly involved in the project, and will not really be members of the core project team, but will be engaged in it for tasks such as data collection. Some SMEs, such as engineers or suppliers, may only need to be consulted at certain points in project research.

## THE PROJECT CHARTER TOOL

The project charter summarizes all of the essential information about a Lean project onto a single page. Many organizations that are engaged in Lean already have a standard project charter form, sometimes referred to as an A3. If this is the case, Lean projects oriented around garbage should simply use the existing system. Otherwise, a project charter should contain the following major elements:

- **Title of your project:** The project should be given a clear title that summarizes the scope.
- **Problem statement:** The problem statement articulates the problem to be addressed simply—in no more than two or three sentences—and explains why the problem is worth addressing. It has an

identifiable defect/gap, specifies where and when the problem occurs, quantifies how large the impact of the problem is, and describes how people know that a problem exists (e.g., a worsening trend, comparison between different areas, comparison against an industry benchmark, etc.). The problem statement should not imply a solution, and should only include a cause if the cause is clear, well understood, and supported by data. If waste costs have been measured, the problem statement should include them.

A problem statement might look something like this: "The scrap rate of nondefective raw materials in the X department at X facility has been 20% for the past three years, compared to an industry standard of 5%. The materials, labor, and disposal costs for this scrap are $X higher annually than the industry standard."

- **Project scope and goal:** The project scope defines the boundaries of the specific project to be addressed within the larger problem described in the problem statement. It identifies where the project starts and ends, what process elements are included, and most importantly, it makes perfectly clear what process elements fall *outside* of the project. *Defining* this at the beginning of a project, and then sticking with that definition, are essential to keep the project from ballooning out of control and becoming victim to "scope creep."

    As a general rule, the scope should be focused as narrowly as possible in order to isolate the problem, limit the variables that can clutter the data, and act as an unambiguous demonstration of the project's effectiveness. Even though the problem statement defines the scope of the *problem* being studied, the *project* scope is usually much narrower. Very often, the problems described in the problem statement impact wide swaths of an organization, but the actual project to work on them needs to be extremely focused, at times even to the point of being an initial pilot project at a single workstation.

    A common cause of failure among Lean projects is inadequate scoping, leading to an unmanageable or immeasurable project. Especially to those new to Lean, the initial project scope needs to be much smaller than might intuitively seem necessary. When developing a project plan, an initial pilot is often the first phase before full implementation, and demonstrates the effectiveness of a solution.

The next phase of implementation can be broader with more confidence, less risk, and greater external support.

A *project goal* articulates what the project intends to accomplish and its expected timeline. Some projects can set specific waste reduction goals at their outsets, especially if potential solutions are already in mind. For other projects, any initial goal projection would only be a guess. Rather than make a true guess, the goal should be left blank until a projection can be made that is driven by some kind of data, however speculative. In general, it is best to underpromise and overdeliver on project goals, especially if a deliberately low-balled project goal already provides adequate justification for pursuing the project. Keep in mind that a project's success is judged against its goals rather than any kind of objective standard, so a high dollar savings that is much lower than the goal projection may still be viewed as a failure by management.

An example project scope following the problem statement above is as follows: "This project will focus on reducing the scrap rate resulting from material handling and storage practices in the receiving department. The majority of scrap occurs in receiving, and this project sets a goal of a 50% reduction in overall scrap by the end of the fiscal year, resulting in an annual savings of $X."

This scope and goal statement could not be written without a deeper understanding of the problem, especially in knowing roughly where and how much scrap is produced in different areas.

- **Project team:** List who will fill each of the project team roles outlined above.
- **Timeline:** The timeline outlines the anticipated completion dates of major project goals or components (milestones) at a very high level. A more detailed project timeline will be drafted as part of the step-by-step implementation goals. For example:
  - September 1–30: Analyze scrap rate by department to verify that the highest loss of materials occurs in receiving.
  - October 15: Build value stream map of existing receiving process.
  - October 22: Perform cause and effect analysis of losses in receiving.
  - November 1: Brainstorm and build future state map for receiving department process and flow.
  - November 5–6: Perform rapid improvement event in receiving.
  - November 6–December 6: Collect follow-up data on scrap rate.

Apart from the timeline proper, be sure to set time aside for follow-up data collection as an essential part of a full project timeline, documenting the true effects of the changes that have taken place.

This project charter should be a living document, and will likely be updated over the project's life as elements change. These charter elements can be collected on a single page, as illustrated below.

The second page of the project charter should be a signature page, with a space for all of the team members and the champion to sign off on the project as described. Although this is more of a symbolic action than a binding agreement, it solidifies everyone's involvement in the project and can be brought out later in the life of a project as a reminder if the champion "forgets" that he or she promised to provide resources and support for this project.

## Project Charter Example

Title: Receiving Scrap Reduction Project

| Problem Statement | Team Members |
|---|---|
| The scrap rate of nondefective raw materials in the X department at X facility has been 20% for the past three years, compared to an industry standard of 5%. The materials, labor, and disposal costs for this scrap are $X higher annually than the industry standard. | Champion:<br>Process owner:<br>Team leader:<br>Team members: |
| **Project Scope and Goals** | **Timeline** |
| This project will focus on reducing the scrap rate resulting from material handling and storage practices in the receiving department. The majority of scrap occurs in receiving, and this project sets a goal of a 50% reduction in overall scrap by the end of the fiscal year, resulting in an annual savings of $X. | September 1–30: Analyze scrap rate by department to verify that the highest loss of materials occurs in receiving.<br>October 15: Build value stream map of existing receiving process.<br>October 22: Perform cause and effect analysis of losses in receiving.<br>November 1: Brainstorm and build future state map for receiving department process and flow.<br>November 5–6: Perform rapid improvement event in receiving.<br>November 6–December 6: Collect follow-up data on scrap rate. |

## VALUE STREAM MAPPING

The process mapping discipline is central to Lean, and can take different forms depending on what information is being captured in the map. In all of its forms, a process map is a graphical tool that clarifies flow by literally drawing a picture of the process. Data from how the process operates are integrated into the picture visually. A sequence of maps makes an objective argument for change by showing different configurations of the same process: how the process currently operates, how it could operate, and explaining the costs and changes necessary to make the improvements. Value stream maps show a multidimensional view of process flow, allowing the reader to see the transformation of the work in progress, the material flow, and the communication between process steps.

The most common type of Lean process map is an expanded flow chart, called a value stream map,[1] which begins with a list of process steps, and then fleshes that basic information out with data about process operations. This may include the distance traveled, scrap rate, touch time, inventory levels, etc. An extensive set of value stream mapping symbols can be used, with many symbols tailored to describe specific Lean improvements. The best resource for exploring these in depth is the book *Learning to See* by the Lean Enterprise Institute. It is beyond the scope of this book to cover this entire vocabulary, but with a few basic symbols, one can map out most processes (Figure 5.1).

The following value stream mapping example shows how some of these icons work together to illustrate the flow of information and material through a process from supplier to customer. Material flows from left to right, and information flows both directions (Figure 5.2).

## CAPTURING GARBAGE ON A VALUE STREAM MAP

Some aspects of garbage and material waste are already captured in a traditional value stream map through the measurement of scrap and defects. Scrap is any raw material or components—other than packaging—that are left over after the process is complete. This can include extra bolts, shavings, scrap cuttings from raw material, material that arrived unusable

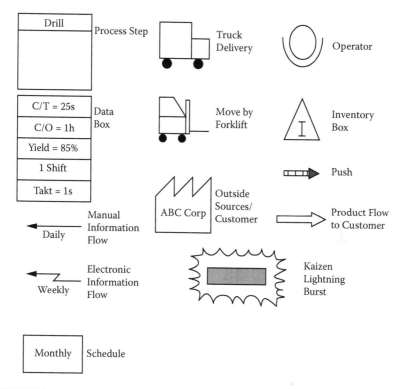

**FIGURE 5.1**

A sampling of common value stream mapping symbols that clarify process flow and the application of Lean principles.

because of supplier problems, expired material, extra unused copies of reports, etc., but *not* packaging. In cases where defective products are themselves the material that become scrap, these two categories measure the same thing. Defective products that are repaired or recovered are not counted as scrap.

## Garbage and Material Waste Data Box

Value stream maps are valuable tools because of their ability to distill data about the operation of processes. These data are usually captured in a data box placed directly below each process step, and display relevant information about it. The following table lists a number of potential metrics to track with respect to garbage and material waste. Few projects would need to track all of this information. In completing a value stream map, track as many of these metrics as makes sense within the context of your project;

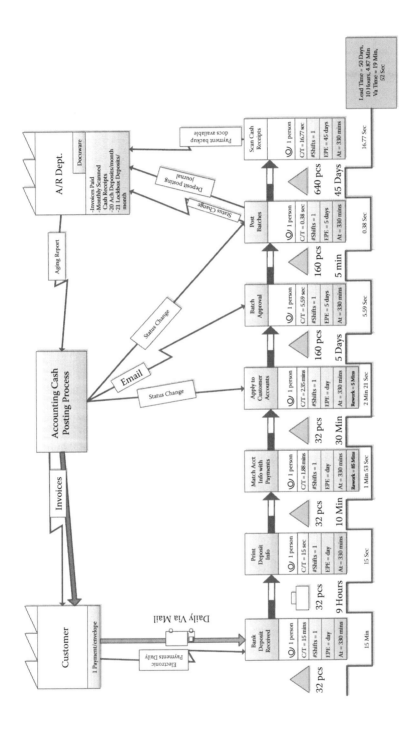

**FIGURE 5.2**
Value Stream Map example.

do not track elements that are outside the scope of your project. These would be measured at each process step delineated on the map.

| Category | Unit |
| --- | --- |
| Total material input | Material type(s) and weights |
| Total material output | Material type(s) and weights |
| Recycled input | Percent or weight by material type |
| Recyclable output | Percent or weight by material type |
| Defects | Percent, weight, or number of units |
| True scrap (excluding defects) | Percent or weight by material type |
| Disposable packaging not recaptured | Percent, weight, or number of units |
| Disposable packaging reused or recycled | Percent, weight, or number of units |
| Reusable packaging | Percent, weight, or number of units |
| Water content of waste | Percent or weight |
| Touch time of packaging, scrap, defects, or other waste | Time |
| Material, worker, or machine wait time | Time |
| Waste removal | Number or frequency of dumpster pulls |

## BUILDING A BASELINE VALUE STREAM MAP

The value stream mapping process begins by building a consensus about the process steps. This may seem like an unusual starting point, but for any process that is performed by multiple people in parallel, there will probably be significant differences in implementation. Even if there is an accepted standard process, the challenge of value stream mapping is to document and agree on the *actual* process that is being performed. It is useful to get as many different perspectives as possible, especially if the process is not already clearly standardized. It can also be useful to build the value stream map in a public space such as a break room where other employees will see it and start asking questions. The full project team usually participates in the initial value stream mapping session:

1. Select the process and decide where it begins and ends.
2. Brainstorm process steps.
   - Write on sticky notes, and post them to butcher paper on a wall. Leave plenty of room, since the results will often be several feet long. Sticky notes are good to use for this because they can easily

be moved around if there are disagreements about the order of steps, or to go back and insert missing steps later.

- Capture what *really* happens, not what *should* happen. To get the best view of a process and not accidentally gloss over steps that are easy to overlook, it can be useful to move backwards through the process step by step from end product back to raw material.

3. Use process mapping symbols to draw the flow of the steps.
   - Arrange the steps in sequence.
   - Check for missing steps or decisions.
   - Number the steps.
4. Identify inputs and outputs for each step. Add the material and information flow into the map (material should flow forward, information should flow backward).
5. Identify key metrics for each step. What should be measured in order to really understand performance?
6. Add a timeline to the process map, divided into value-added and non–value-added pieces.
7. Review and finalize the process map.
8. Date the final drawing.

Any initial estimates that inform this map must later be validated with real observations. Usually, the observations come from physically walking through the process and collecting measurement data of how long process steps take, how long materials wait between steps, or how much raw material is used. The data collection process can be extensive. At times, this process data may be collected from historical documentation of process performance. This initial map is often cleaned up into a more presentable form using a software program such as Visio (which has a value stream mapping plug-in available), but hand-drawn maps have the same functionality.

The product of this initial value stream mapping session is the baseline map, showing the process before any improvements have been made. The baseline map is also sometimes referred to as the "as was" map to emphasize the fact that it documents past practices that are in the process of being changed. Whatever steps and measurements are agreed upon as accurately describing the baseline should be considered the standard operations for the process until improvements are made to it.

It is extremely important to resist the urge to start making improvements in the process at this point. The baseline map should come as close as possible to documenting the baseline process without influencing it or even fixing obvious problems. Problem solving will only come after the existing process is fully understood. The effectiveness of improvements in the final project is only judged against the baseline. The results of any undocumented improvements will not be measured as part of the final improvement, and undocumented improvements will not themselves be critically evaluated for effectiveness, since they cannot be effectively compared to the original process.

## MAPPING TIME

Time is usually measured with a serrated timeline below the process steps, which divides time into value-added and non–value-added portions. Non–value-added time includes the time spent on rework, redundant process steps, waiting time, or other uses of time that do not contribute to customer-driven requirements. These portions are then totaled up to produce a picture of the entire process's cycle time and the relative breakdown of value-added vs. non–value-added components (Figure 5.3).

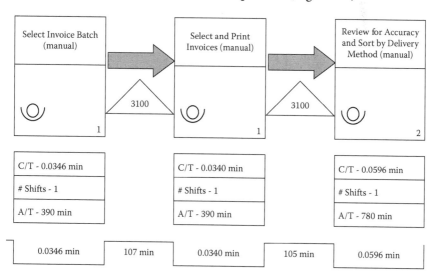

**FIGURE 5.3**

Value stream mapping (VSM) detail showing data boxes with timeline running underneath. Upper values are non–value-added (NVA) time, and lower values are VA time. Note that large inventory queues result in a very high proportion of NVA time, making the process itself quite lengthy even though the work time per unit is low.

A good way to start analyzing time is to divide it into wait time and touch time. All wait time is non-value added, and touch time at least *provisionally* represents value-added activities since it represents activities in the baseline process. Reductions in touch time will be felt by the organization by freeing up staff for other duties, and reductions in wait time will be felt by the customers as a quicker response from the organization. It is ultimately important to measure all kinds of time in the same units so that they can be compared directly (for example, it might often seem more appropriate to collect touch time in minutes or seconds and wait time in days, but they should both be notated in minutes for the purposes of the map).

There is a debate within the Lean world about how to account for reductions in labor time associated with process improvement. Reductions in the labor standard clearly increase a worker's capacity by allowing more material to flow through his or her process in the same amount of time. This improvement in productivity translates into greater profitability for the process as a whole. However, reductions in labor time are not—in themselves—hard savings. If a worker's time costs a company $25/hour, streamlining three hours of work out of that person's process per week does not result in a real $75 savings, because the worker's full salary will still be paid. The real value comes from calculating the worth of the value-added time. A worker's time is usually measured in terms of how much the time *costs* the company, but a better way to approach it from a Lean perspective is to ask how much that worker's time *earns* the company when he or she is performing value-added activities. At $25/hour, that worker costs the company roughly $0.40/minute for time spent waiting in line or doing other non–value-added activities, but that same time is probably *worth* several dollars a minute when he or she is working on value-added activities. Calculating the cost of every employee's time per minute can still be a useful exercise to help begin to see the impact of wasted time.

Labor reductions can only be measured as direct savings when they either reduce overtime pay or free up internal capacity that would allow the business to avoid hiring additional people. Many Lean projects include dollars from labor savings in their project totals, but this should only be done as a measure of potential capacity increase and not as hard savings. Anything listed as a hard savings runs the risk of being removed from a department's budget in subsequent fiscal years.

## MAPPING IMPROVEMENTS

Very often, project teams will create three maps: a baseline, a transitional map showing the opportunities for improvement identified by the team, and then a future state map showing the performance of the process after the changes have been put into place. Transitional maps can be thought of as rough drafts, or as maps that capture partial, in-progress improvements leading to a final state. For very long or complex processes, several transitional maps may be employed to show phases in an extended improvement plan. The future state map is the blueprint for the improvements that the team will make to the process. It uses the data on current operations to identify areas for improvement and project the impact of any changes.

In looking for improvement opportunities or areas of waste, analyze the baseline map using the full palette of Lean tools:

- Which processes do not meet takt time?
- Where could level loading be implemented to balance steps?
- How could processes be simplified, combined, or eliminated to improve the flow?
- Could setup time be reduced through breakaway fixtures, multiuse fixtures, or kitting?
- Where could continuous flow or single-piece flow be introduced?
- Could flow and inventory be improved through the use of kanbans or just in time?
- Would cross-training improve productivity?
- How could inventory between processes be reduced through first in, first out (FIFO), supermarkets, or milk runs?
- How could transportation time be reduced?
- How could product flow throughout the facility be improved?
- Would 5S benefit any particular areas?
- Could inspection times be reduced using go/no-go gauges?

These are examples of some of the more common Lean tools that can help identify opportunities for improvement to integrate into the future state design. The process of creating a future state map is often more open-ended and creative than the baseline map, and can involve looking for opportunities for simplification that fall outside of conventional assumptions. In order to produce meaningful change, the solutions that inform the future state map should address the root causes of problems, rather than just treating problems that are really symptoms. There are great Lean tools for this kind of assessment as well, such as the five whys exercise or the use of cause and effect diagrams.

## PROCESS MAPPING PRESENTATIONS

Most mapping projects will have five essential components in terms of presentation. The purpose of this presentation is to make an argument convincing management that the improvement under consideration produces a meaningful impact.

- **Project charter:** What is the problem? What is the intended scope of work within that problem, and what is the goal for improvement? Who is going to be involved? What is the projected timeline for completion?
- **Baseline map:** This map captures data on how the process operated before any improvements were made to it. This may well include unflattering facts about operations, and should be used to demonstrate the causes of the problem identified as the focus of the project. Often a baseline map will be accompanied by a series of charts or other supporting data.
- **Future state map:** This map envisions the redesigned process after applying Lean principles for waste elimination and minimization. Process operation data in this map should reflect estimated changes based on how the known operations would be impacted by changes.
- **Action plan:** This is a detailed list of the steps that are required in order to make the improvements outlined in the future state map, including all anticipated costs for implementation. The action plan should cover every gap between the baseline and future state process.
- **Return on investment (ROI):** The ROI is a comparison of the baseline and future state maps showing the amount of waste that can be eliminated through this project, including both wasted resources (time, materials, equipment, etc.) and the value of those resources to the organization. This specifically puts a dollar value on the waste as compared to the savings. In contrast, the action plan puts a dollar

value on the cost of fixing the problem. ROIs usually express the payback period for investing in a change in terms of how many years the change will take to pay for itself. As a general rule, ROIs of two years or less are considered highly desirable, and demonstrating this as a payback period will almost certainly win approval for a project.

Two other types of process maps are outlined in other chapters of this book, applicable as tools to measure specific wastes. The spaghetti diagram is introduced in Chapter 8 to measure waste from transportation, and the swim lane map is introduced in Chapter 11 to measure waste and delays involved in the handoff of material (something that is particularly applicable to paper processes). Many other types of maps exist and may be applicable to individual projects, and all types of maps may be fleshed out with data boxes.

## MAKING IMPROVEMENTS THAT STICK

There are several factors that together impact the ability of improvements to remain in place, rather than reverting back to the original process. These factors include the design of the improvement, the level of management support, and the engagement/transparency of the system. Making a successful improvement effort involves as much psychology as it does organization, and it is reliant on making the system as prone to success as possible. Figure 5.4 shows a number of different approaches that organizations attempt to implement in order to fix problems. Notice that awareness and training are at the bottom of this figure, having the least tangible effect, and yet many organizations focus on these as their primary efforts to drive change. These may be good first steps, but only as they are tied to real-world improvements that are driven by process simplification, error-proofing, or automation. These more effective improvement approaches have a greater impact because they do not rely on behavioral change, and instead revolve around making the process itself inherently prone to success.

As a guiding principle of Lean, management is *always* responsible for the performance of processes, so the role of management or the project champion is especially important in ensuring that the improvements will be maintained. In part, this is seen in management demanding performance

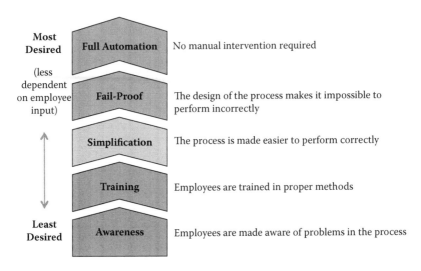

**FIGURE 5.4**
Increasingly effective methods at improving the performance of processes. Notice that the process is made more autonomous and self-correcting as one moves up.

metrics from the process, and reviewing them over time to observe that the improved metrics remain in place as the new norm. If the process begins to drift back toward its older performance standards, management can step in and ask for a review or hold people accountable. Management must also allocate adequate resources to the project group to maintain their improvements, and a failure to maintain improvements may really represent a failure on the part of management to provide support.

One of the simplest tools for helping maintain improvements is to simply publicize data on how the process is performing. This often takes the form of a dashboard or periodic report that is sent out to everyone involved in the process, or it may be a visual display board in a prominent location with performance metrics. These tools not only give people who operate the process the data they need to understand their performance, but it also provides accountability through transparency. By itself awareness of process performance is not adequate to drive improvements, but it is an important component necessary for them to succeed.

# ENDNOTE

1. Value stream mapping is a highly developed discipline. The best available resource on value stream mapping is *Learning to See*, published by the Lean Enterprise Institute.

# 6

## Effective Recycling Programs

Purchasing recycled products and producing recyclable ones are both important parts of landfill diversion efforts, and there are many ways in which the act of recycling can fit into the larger material flow of a business. In this sense, recycling has both a material waste and a garbage component that are equally important. The classic conception of recycling as a closed-loop system where a postconsumer product is remanufactured back into itself at the end of its life cycle is atypical. Waste products are instead more commonly manufactured into different products, especially where a recycled material is incorporated into the manufacture of a more complex new product.

## DESIGN FOR RECYCLABILITY

Recycling is something of a chicken-and-egg problem. It is equally important to have both manufacturers who produce products that are capable of being recycled and manufacturers who purchase recyclable waste as raw materials. In order to support this, it is also important for companies to set purchasing policies that give preference to recyclable or recycled products whenever available.

In order for products to be recyclable at their end of life, it is important to engineer them with this intention from the design stage. This design issue includes minimizing mixed material products that have to be deconstructed before their components can be recycled, as well as minimizing products that effectively cannot be deconstructed because their component materials have been fused or combined together into a composite material. This is especially true of products that include different types of materials layered together (such as plastic on metal). With the exception of especially valuable materials such as copper or titanium, the value of most recyclable materials is low enough that *any* investment of time required to deconstruct a product will likely result in it not being recycled by businesses.

The purchasing and production sides of a single industry can sometimes be aligned into a closed-loop recycling program. Where a closed-loop recycling program exists, the products manufactured by a business are converted directly back into their own raw materials at the end of their life cycle. This is a difficult challenge, and requires the adoption of an industry standard for material composition that is agreed upon between major manufacturers. This approach makes the most sense for products made from a single material (such as aluminum cans) that can be easily rendered back into their original form. This approach requires very rigorous design for anything more complex than a simple metal can. Even with aluminum drink cans, it is worth noting that the design of the can was reengineered over a process of over 30 years to create an industry standard can entirely from aluminum that could be recycled. Prior to 1980, cans in the United States were made of several different types of metals, and continue to be so in some parts of the world.[1]

The Atlanta-based company Interface carpeting is one business that practices closed-loop recycling exceptionally well.[2] Within the framework of recycling, its carpeting is not *sold* to clients—instead it is *leased* to them. Interface positions itself more as a service provider than as a product industry. Used carpet is recollected at the end of leases, and reprocessed into new carpeting. This results in very low material waste, increased customer retention, and significantly lower material costs since they effectively do not have to pay for their raw materials after material gets into the system. However, Interface's recycling process only works with its own carpeting, because it has been specially designed to be recycled. This is one way to avoid the requirement for an industry standard, but it is not optimal, and restricts Interface from working with materials from other companies (Figure 6.1).

## THE PSYCHOLOGY OF EFFECTIVE RECYCLING

The low dollar value of many recyclable materials is a barrier to recycling, and means that materials need to be very efficiently handled in order for the entire recycling operation to be cost-effective. Above all, the

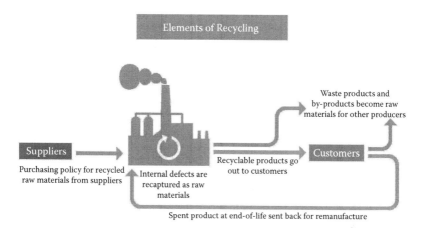

**FIGURE 6.1**

Major points in the material process where recycling should be considered from a design and supply perspective.

requirement of excess labor from the workforce to implement recycling programs should be minimized, because this will lead to resistance from all levels of management, who will correctly see time spent on recycling as time taken directly away from productivity.

The most important element to consider in establishing an effective recycling program is the ease with which staff have access to it. The act of recycling needs to be as easy as—or easier than—the act of conventional waste disposal. Poorly designed recycling programs often require people to make a special effort to recycle, usually through improperly located recycling containers. If *any* kind of extra effort is required to recycle—even a single step—participation in the program will be severely impacted, and the only people who will participate are those who see recycling as intrinsically valuable or those who fear repercussions for noncompliance.

As a bottom line, if workers perceive recycling as something that pulls them away from the tasks that their supervisors use to evaluate their performances, they will not participate. In this sense, a well-designed recycling program should increase productive time with respect to waste handling, and should be one of the metrics that drive employee performance evaluations.

Effectively designed recycling programs do the following:

- **Place recycling containers at the locations where the recyclable waste is being produced.** For example, the recycling container for soda bottles does not belong next to the drink machine. The machine is where the *full drinks* are produced, not where *empty bottles* are produced. The contents of the trash cans around the facility can be analyzed to learn where people actually dispose of the most bottles, or the process of consuming drinks can be observed and mapped to track the flow of bottles. These will indicate where to place the bottle recycling stations so that they will be maximally convenient and require people to carry empty containers the shortest possible distance (hint: this will probably be in the break room).

- **Design recycling containers to accept only the proper type of waste in order to minimize accidental contamination.** This is often done by designing access to those containers through openings shaped like the desired material, such as a thin slot for paper, a small round hole for bottles or cans, or a screen for sand. In Lean terms, this is known as a go/no-go gauge, a process control that uses the shape of a container or fixture to prevent workers from placing an incorrect object in that container (Figure 6.2).
- **Clearly label recycling containers with the type of material that they are intended for.** If workers are confused about what container a particular material is supposed to go in, they will either potentially contaminate the recycling stream or throw the material away rather than taking the time to research its proper disposition.

Beyond these steps that simplify and encourage recycling, it can also be useful to remind people of the ultimate destination of the nonrecycled garbage as well. For example, recycling containers almost always carry a

**FIGURE 6.2**
Series of recycling bins with holes shaped like the materials they are designed to collect.

label for their contents (e.g., "glass"), and the regular garbage can likewise be labeled "landfill."

## SHARING THE REWARDS OF RECYCLING

There is an incorrect tendency among many environmental advocates to assume that recycling is something that most people see as inherently valuable, and that everyone would participate in it altruistically if it were convenient. Because of this, problems of *access* are typically treated with more weight than problems of basic *motivation*. After all, household recycling is something that people engage in with no expectation of any kind of compensation. In fact, their waste bills are slightly higher as a result of having access to this program, and their only direct reward is a feeling of good environmental stewardship.[3]

In the world of business recycling, a feeling of good stewardship is usually not an adequate motivator. Although some businesses hold material conservation as a core value, most do not. For business recycling programs to succeed, it is important to tie some kind of direct benefit of recycling back to the people who handle the material. Participation in recycling should be something that supervisors track as part of employee evaluation, so that recycling is not something that employees see as being *outside* of their regular job duties.

Another good way to drive recycling is to have the profits from recycling go into the bonus fund for employees, and to make the value of recyclable materials very clear to employees in these personal terms. This creates a tangible motivation for participation in recycling. Sometimes this can even be a separate bonus, listed as an itemized amount in the regular bonus or paycheck. This approach helps not only motivate individual employee behavior, but accomplishes the larger cultural goal of encouraging the workforce to self-police on recycling, since everyone's behavior impacts the bottom line.

## WASTE SORTING AND SEGREGATION

Proper waste segregation is the first challenge in creating a successful program, both in getting all of the recyclable materials into the recycling stream and in preventing contamination of that stream from undesirable waste. This is similar to the challenge faced in dealing with regulated waste, but without regulated waste's threat as a contaminant in the conventional waste stream.

At some point in the production of waste, the recyclable materials will have to be effectively segregated from the rest of the waste stream. In general, the earlier in the process this sorting occurs, the less overall labor the sorting process will take. Less sorting and handling is involved if material is segregated earlier. Thinking of this in Lean terms, rework is always more expensive than doing a job correctly the first time. Labor can be optimized either in terms of time or in terms of cost, and multiple perspectives on maximum value should be considered. Two ways to approach optimizing the labor involved in recycling are

1. The minimum absolute amount of time should be spent handling the material, regardless of who handles it.
2. The employees whose productive time is of the lowest productive value to the business should sort the material. The value of time takes into consideration not only the *cost* of time through salary level, but also the *income* that that employee generates while productive. By this hard logic, the value of custodial time is much lower than the value of production workers, who are creating the product that earns a profit. This is not meant to belittle the value of custodial work, but to address the primary concern of management in tying up productive workers with tasks other than production.

Ideally, the people producing the waste should have the facilities to at least partially segregate the waste themselves as it is being produced, so that it will not have to be handled again as part of any sorting process, and this process should be no more labor-intensive than using the conventional waste stream.

Virtually all recycling programs of disposable materials live or die by the custodial staff, since the responsibility for handling the waste often falls to them. In many cases, recycling programs fail because they create an impossible workload for the custodial staff. Custodians might have to make two separate sweeps of a facility to collect refuse instead of just one, but are not given any more resources or time to collect it. In this case, recyclable materials may quietly be placed in the regular garbage. If your garbage audit uncovers that this is happening, the first question to answer is whether or not the custodial staff has been given adequate resources to cope with the workload. Custodial capacity may need to be adjusted to account for recycling.

Since well-implemented recycling programs reduce the volume of conventional waste by diverting a portion of it away, in some cases this can reduce collection needs for conventional waste. For example, if conventional waste collection can be conducted more infrequently due to recycling, then the labor time necessary for recycling collection can naturally be freed up. Whether it is recycled or landfilled, the same amount of waste is generated and handled, so the labor costs should be equivalent if managed well.

## MAKING GARBAGE TRANSPARENT AND ACCOUNTABLE

While recycling itself is something that is generally considered desirable, the nontransparency of the waste stream means that there is little or no accountability in most cases for *not* recycling. Recyclable materials in the conventional waste stream are usually invisible, and when they are visible, they are usually untraceable.

One of the desired outcomes of a Lean waste stream is accountability with respect to waste, by making the waste stream transparent. In part, this goal is accomplished through the fact that people are aware that the garbage is being analyzed. The associated challenge is to analyze the garbage without making people feel like they are living under surveillance. If at all possible, garbage transparency should not produce an adversarial relationship.

In some instances, materials in the garbage can be traced back to a clear source. One good example of this is cardboard, which can often be traced back to its source by means of its shipping label. As part of an ongoing garbage monitoring process, this makes cardboard one of the easiest materials to address, and it can be used as an example to emphasize the transparency of the garbage. Even in situations where there are no formal consequences for not recycling, the garbage auditors can do something like clipping the shipping label from every box found in the trash, and sending a form email on waste prevention efforts to whoever received the package. These emails can be made slightly public by cc'ing everyone from a day's findings on a single email. This practice opens an interesting dialogue about garbage, deeply offending some people and positively engaging others.

---

## POSTCONSUMER RECYCLING VS. MATERIAL RECAPTURE

Especially in manufacturing, internal systems are often set in place to recapture work in progress when errors are made in the manufacturing process. For example, a pet food facility that produces dry dog food might have a hopper set up on its production line so that any food that is improperly bagged can be fed easily back into the production line. This should not be thought of as recycling, but as waste prevention. If possible, this kind of waste prevention offers the opportunity to recapture defects directly as raw material. This distinction between preconsumer and postconsumer waste is important in evaluating recycled products.

In cases where recycling programs handle disposable materials such as packaging, recycling should be thought of as a transitional approach to waste diversion at best. Although it successfully diverts the material from the landfill, the program itself can be costly to operate, and is a kind of stopgap that can eventually be scaled back as material reuse and reduction programs eliminate the sources of the recyclable material. In cases where recycling recaptures the scrapped raw material or postconsumer product, the recycling effort is an integral part of ongoing operations to recover production waste.

## ENDNOTES

1. Steel and composite beverage cans have been popular since the 1930s, and the industry gradually shifted from the 1950s to the 1980s to a new standard of all-aluminum cans. To read more about the history of drink cans, visit http://www.aluminum.org/Content/NavigationMenu/TheIndustry/PackagingConsumerProductMarket/Can/default.htm or http://en.wikipedia.org/wiki/Beverage_can.

2. Interface carpeting has also pioneered a number of other advances to reduce waste from the carpet industry. Most notably, Interface pioneered the concept of carpet tiles—where flooring is covered with small, independent squares of carpeting instead of a single sheet—allowing for the strategic replacement of tiles instead of replacing the entire floor. Learn more about Interface's programs at http://www.interfaceglobal.com.

3. The Environmental Protection Agency (EPA) provides a useful reference for estimating the cost of household recycling (http://www.epa.gov/osw/conserve/tools/localgov/economics/collection.htm). Household recycling only has a direct benefit to the participants in cases where their contributions are measured and credited, or if the cost to dispose of garbage is calculated by weight. The Recyclebank company is an excellent model of this practice (https://www.recyclebank.com/).

# 7

# *Composting Programs and Organics*

Composting is the natural end of the life cycle for organic materials, and can be a good way to dispose of biodegradable waste, usually food waste or plant matter. The only distinction between recycling and composting is that composting uses biological processes instead of mechanical ones, and produces an end product that may be most suited to biological uses. It follows that some kinds of organic waste can be either composted or recycled, depending on which approach produces the most useful output material with customer demand.

It is far better from a Lean perspective to avoid the production of compostable material than to reprocess it after it has been created. As with recycling, the challenge in composting is in avoiding excess labor. Unlike recycling, there is often an extra barrier to composting in finding external customers willing to accept the waste material at all, or in finding a meaningful use for the composted material after it has been broken down. Some municipalities have citywide compost facilities, but these usually only accept plant/landscaping waste, and may not accept even that from businesses. The fact that compostable materials begin to decompose almost as soon as they are generated adds an additional challenge to their reclamation.

## FOOD WASTE

According to the Environmental Protection Agency (EPA), food leftovers are the single largest component of the waste stream by weight in the United States, accounting for more than 25% of the food we prepare (roughly 96 billion pounds per year).[1] Food waste is complex and difficult to address for many reasons, mostly driven by the problems of its perishability.

- **Human consumption:** Food waste that is simply left over but is otherwise edible by humans can be donated, and many shelters and other organizations are happy to coordinate the pickup of leftover food. One of the barriers to donating leftover food for human consumption is the fear that anyone who might be sickened by the food could sue. Legally, this fear can be answered with a liability waiver—but for many restaurants, the question goes beyond legality to the larger problem of reputation, and this risk may still be a deterrent to human consumption of leftovers.
- **Animal consumption:** Food scraps, either from the preparation of food or as scrapings from plates, can sometimes be donated for animal consumption, although the same fears with human consumption often apply here as well. Contaminants in this food supply—even things like toothpicks or napkins—can render this food inedible to animals.
- **Industrial uses:** Depending on the contents of food waste, it can potentially be processed into biodiesel or have its fat content rendered for industrial uses.
- **Compost:** Practically all food waste can be composted if facilities are available. Unfortunately, it can be difficult to find compost facilities that are willing to accept the volumes of food waste produced by any significant operation.
- **Digestion:** An alternative to composting food waste is the installation of a food digester that uses microbes to quickly break down food waste into gray water. This water can then be land-applied as a liquid fertilizer, since it is very nutrient rich. Digesters capable of handling large volumes of food waste are available commercially.[2]

Source reduction strategies for reducing food waste volumes are much better than waste handling solutions with respect to food.[3] Portion control is an important aspect of this with customers, and strategies such as the removal of trays from all-you-can-eat buffets will help prevent people from taking significantly more than they can eat, while not actually limiting the amount that they may take in total.

## PLANT WASTE FROM LANDSCAPING

Nonfood plant waste is the other major category of compostable materials—often referred to as green waste. The most commonly produced kind of plant waste is landscaping waste, since organizations of virtually all types maintain landscaping. From the perspective of organizations, it can be difficult to control the disposition of this kind of waste because a great deal of landscaping maintenance is contracted out rather than being handled internally by permanent staff.

Like other waste materials, the garbage produced by landscaping ties back into the efficiency of the landscaping design itself. Landscapes that are appropriately designed around the local climate and environment require minimal water, fertilizer, and maintenance, and produce less waste material. If landscaping is truly designed along principles taken from the local environment, it may need no additional water. As a point of comparison, it is worth noting that most truly natural landscapes also produce no material waste that leaves the space, with dead plant matter decomposing back into the soil in the same environment.

Appropriate landscaping is dependent on both regional and micro climates, and should be designed in consultation with local expert gardeners. As a set of general principles:

- Mulch any exposed soil to prevent weeds and evaporative water loss.
- Utilize regionally native plant species with an emphasis on low-water-use/drought-tolerant plants, with xeriscaping as a potential goal. This is especially important for tree plantings.
  - Keep in mind that drought-tolerant and native plants can be two different categories. Drought-tolerant plants may be adapted to survive well in a region with little rainfall, but nonnative drought-tolerant plants may not interact properly with local pollinators and other species. This can result in both invasive species problems as nonnative plants escape domestication and a lack of support for local animal and insect species that rely on plants to survive.
- If automated sprinkler systems are used, make sure that they are regulated as efficiently as possible to minimize water waste, including soil moisture sensors.
- Use permanent plants (hardscaping) that do not have to be changed out with new plantings seasonally. A hardscaped bed can be designed

so that plants that are active in different seasons are coplanted. For example, spring, summer, and fall blooming perennials can grow together, and will naturally change the appearance of the landscape seasonally.

- Minimize mown grass with an emphasis on natural areas that can be allowed to grow unhindered. Unmown natural areas can be developed under the canopy of heavily shading trees as well as full sun using different kinds of groundcover. These can feature groundcover plants, grass species that only grow a few inches tall, landscaping that revolves around rockwork, or attractive "no mow" zones can be created for large grassy areas that do not have heavy pedestrian traffic. A no-mow zone reverts to a quasi-natural state, giving at least the appearance of wilderness grassland. No-mow zones can sometimes be branded as wildflower meadows or bird sanctuaries to increase their public relations value, and can represent a very visible environmental improvement to a landscape.

- Design plantings to help with energy management wherever possible. Trees on the south and west sides of facilities can help shade them in the summer months, and windbreak plantings on the north can minimize energy loss in the winter. Climbing ivy on any wall tends to regulate temperature variation as well, through both shading and respiration. Where structures will tolerate the weight, green roofs also provide good insulation.

- Avoid the use of chemical fertilizers as much as possible. Particularly when applied to grass, one of the major effects of these is simply a need to mow more frequently.

- Container plantings can be integrated into virtually any space, and the introduction of container plantings inside facilities has the added benefit of increasing interior air quality.

## MUNICIPAL COMPOST YARDS

Some municipalities set up dedicated compost facilities to deal with citywide green waste. The cost to operate these facilities is often more than offset by the direct savings in tipping fees for landfilling the material. Additionally, these facilities produce finished compost several times a year, which can be given back to community residents, or even potentially sold if properly processed.

In order to make a municipal compost facility most effective, its establishment should be paired with the adoption of a city ordinance prohibiting green waste from being placed in the conventional garbage. This segregation requirement does not place any additional burden on the citizens, and ensures that the city has an enforceable mechanism to maximize the benefits of the compost facility. The ordinance should allow the issuance of citations for contamination of either the conventional waste stream with significant green waste or vice versa.

---

## PLANT WASTE FROM OPERATIONS

Some industries that actually process plant matter in their operations produce plant waste as a direct by-product of operations. For many operations, waste wood will be the most common plant product in the waste stream. Waste wood, such as pallets, can be chipped and composted, or even sold as wood chip mulch or for other processes where wood by-products are needed. As a caution on this practice, mulch from chipped pallets needs to be tested for chemical residues before it is applied to any landscaping bearing edible plants or used in places where people will come into contact with it. Composting waste wood from packaging is not advised in most industrial operations, because the wood is often treated with chemicals to prevent decomposition. Additionally, many wood products—such as composition board or plywood—contain glue and other contaminants that may make them unsuitable for composting. Apart from processed wood, industries that process plant matter into products increasingly have options to use the organic by-products of their operations, such as energy production from waste organics through biogas generation.

### Wood and Cardboard Recycling Example—V&M Manufacturing

V&M Manufacturing makes heavy oil field equipment, much of which is shipped in and out of their facilities by rail on pallets or larger wooden supports. Prior to improvements, wood waste from their processes was being sent to the landfill even though it was being segregated into a special dumpster due to restrictions from the waste management company on accepting wood in the conventional waste stream.

In this project, V&M established a relationship with a nearby company that makes wood pellets to burn in wood-burning stoves. Since the wood waste was being segregated anyway, this improvement did not require a major change in waste handling practices for V&M. The neighboring company agreed to pick up its wood waste at no cost for use as a raw material in its process. Instead of a dumpster to collect the wood, a trailer truck was parked in the same place, which was emptied when full. Approximately 245 tons of wood is diverted from the landfill through this improvement annually.

In addition to the reprocessing of wood waste, V&M also leased a cardboard baler as part of the same project, capturing 56 tons of cardboard in the first year, at a combined savings of $19,400 in lower disposal costs (Figures 7.1 and 7.2).

Beyond these efforts, V&M used the information gathered in its examination of the garbage to standardize and control other aspects of its waste stream as well, installing clear signage specifying where each type of waste was to be disposed of (Figure 7.3).

**FIGURE 7.1**
Wood waste dumpster before improvement.

**FIGURE 7.2**
Wood waste collection truck after improvement.

**FIGURE 7.3**
Signage on dumpster clarifying exactly what should and should not be placed within it.

## BIODEGRADABLE MATERIALS

An increasing range of products advertise themselves as being biodegradable and compostable. This includes items such as compostable plastics that will only compost under very specific conditions. If these materials are landfilled, the fact that they are biodegradable is essentially irrelevant, since landfilled materials will theoretically be sequestered away for the remainder of human history and are disposed of in ways that inhibit decomposition as much as possible.

Outside of their compostability, biodegradable products (such as compostable plastics) are more likely than their noncompostable counterparts to be manufactured using sustainable materials, such as plant-based plastics. Thus, although there is chemically no difference between plant-based and petroleum-based plastic, there may well be other environmental benefits that extend out beyond the disposal of the product when the whole life cycle is analyzed.

If biodegradable products are available at comparable costs and functionality to nonbiodegradable ones, they may be embraced and disposed of through compost systems as a waste reduction effort. However, this may or may not prove to be cost-effective since the additional cost associated with purchasing a compostable product may quickly outweigh any cost benefit from waste reduction. As with problems in recycling, a much greater improvement can be made by focusing on source reductions.

## EDIBLE/CONSUMABLE LANDSCAPING

In some cases, the landscaping can be designed to play a role in the business that it supports. Ideally, the landscaping can produce some of the materials on site that the business consumes. This can not only be beautiful, but also integrate the labor-intensive job of maintaining landscaping into the business model, even if this contribution is really only a gesture.

A great example of this idea in action can be found in Enrique's Mexican Restaurant in Ponca City, Oklahoma. The restaurant is located within a small airport, and although there is very little landscape space for the restaurant to work with, the available planting has been done in edible peppers, which are then used by the restaurant as they mature. While the limited green space provides only a small percentage of the produce needed by the restaurant, it makes the landscaping interesting, rather than just being an irrelevant ornament (Figure 7.4).

As a broader example of edible landscaping, Falling Fruit (http://www.fallingfruit.org) demonstrates the power of mapping the edible landscape from a material exchange perspective. This organization's goal is to create a map of urban food production by mapping out the world's urban

**FIGURE 7.4**
Landscaping pepper plants around Enrique's.

fruit- and nut-bearing trees. This map is then used to align these food resources that go unharvested with organizations collecting food. Like industrial material exchange programs discussed in Chapter 9, Falling Fruit makes the consumption of waste material much more likely by making it visible to people who could potentially use it.

## ENDNOTES

1. See http://www.epa.gov/wastes/conserve/foodwaste/tools/foodcost.pdf.
2. Learn more about the ORCA digester at www.totallygreen.com.
3. The EPA has established a hierarchy of food waste recovery strategies that echoes the points written here: http://www.epa.gov/wastes/conserve/foodwaste/tools/foodcost.pdf.

# 8

## Transportation and Storage of Garbage

For many companies, the labor costs associated with storing, handling, and internally transporting their own waste can easily exceed the fees from contracting with an external waste hauling service to remove it from the premises. The reduction of internal waste handling should be a priority in waste reduction efforts. This is done with the primary intention of minimizing internal labor costs by only pulling staff away from other duties to transport waste the shortest distance in the shortest time. Minimizing the *external* waste handling costs from waste hauling services is really a separate goal from this, and is tied to reducing the actual volume of the garbage being removed.

### PROCESS IMPROVEMENT AND SPAGHETTI DIAGRAMS

The most powerful Lean tool for the evaluation of material transportation is a mapping tool that is usually referred to as a spaghetti diagram. The spaghetti diagram tracks the course that workers or material follow in a process, plotting their paths on an aerial map or floor plan. Spaghetti diagrams can be used to plot movement within a facility, or can be used on a larger scale to plot out vehicle routes on a roadmap. Data for spaghetti diagrams should be collected directly, by walking or documenting real movement. The spaghetti diagram takes its name from the fact that the complicated path on the final product often ends up looking like a plate of spaghetti.

Spaghetti diagrams concretely visualize process flow, and allow the direct comparison of before and after pictures of a process. Most importantly, spaghetti diagrams show *only* the transportation involved in a process, so a person evaluating a spaghetti diagram can evaluate the waste of transportation independently from actions in the process steps themselves.

The following examples show a set of spaghetti diagrams documenting movement in a workspace before and after improvements were made to the layout of that space. This spaghetti map example is from a manufacturing facility. The data were collected for this map by setting up a camera to film a worker at a workstation during the entire process of producing one unit. Improvements were then made based on patterns of movement, focused particularly on shifting equipment to shorten the most frequently taken paths.

Some spaghetti diagrams indicate the number of repetitions of a path, although this example does not. In this example, it is impossible to count how many cycles were completed between the door and the grinder because the lines blur together. Although there are several changes between the before and after maps, the most visible changes are the movement of the grinder to eliminate the most significant source of motion and the installation of an automatic loader. Knowing both the number of repetitions and the distance along the paths would allow the transportation savings to be calculated. Including a scale in the map can make this measurable.

As with other maps, spaghetti diagrams can be beautifully drawn with computer software, and can even be animated as part of presentations to management, but pencil or whiteboard drawings have exactly the same functionality and are much easier to make quickly (Figures 8.1 and 8.2).

For vehicle movement, GPS data can be used to track precise movement. GPS data can also provide a great deal of other data on vehicle usage and performance if desired. A spaghetti diagram capturing vehicle routing can follow the model outlined above, where the precise path of a vehicle is shown as it moves through an area, or a higher-level view can assess regional transportation issues (see Figure 8.3).

If transportation waste is suspected to be a problem, a spaghetti diagram should be made for the entire process, rather than just mapping out the part of the process after material becomes garbage. This will help identify upstream transportation that is also associated with the material, and help calculate the true labor impact of materials reductions.

Spaghetti diagrams are often completed to collect data that will suggest areas for improvement for rapid improvement events focused on process flow, and will work alongside value stream maps. A single process may be mapped in several different dimensions using different mapping tools.

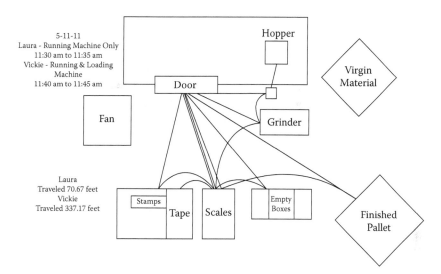

**FIGURE 8.1**
Before map.

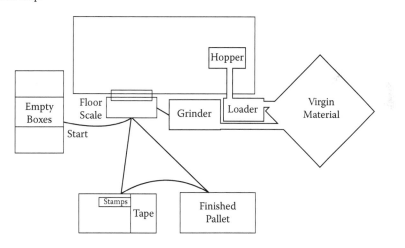

**FIGURE 8.2**
After map.

## GARBAGE COMPACTION

Beyond the labor associated with waste handling, one of the principal problems associated with the transportation of garbage is the fact that a great deal of garbage's volume is usually air. A "full" container that cannot hold more waste might in reality be mostly empty, but the space is unavailable.

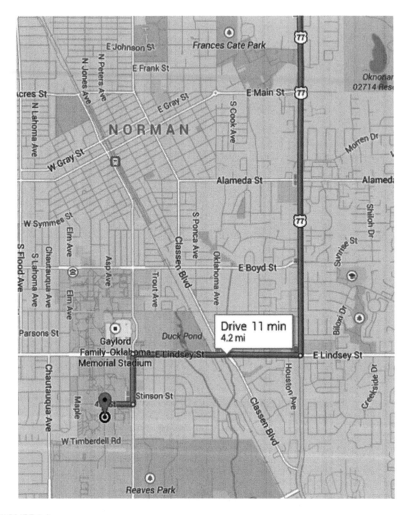

**FIGURE 8.3**

A spaghetti diagram for vehicle routing can be made simply using Google Maps. This tool also can be used to optimize a route that contains multiple destinations.

Packaging and shipping containers are often the culprits of this problem, since their functionality is to protect their interior space from collapse. In Figures 8.4 and 8.5, a dumpster from an office building is shown before and after having all of the cardboard boxes removed. The before picture is overflowing, while the after picture is slightly over half full. A cardboard recycling container sits next to this dumpster.

The installation of a trash compactor has a relatively high up-front cost, but if used appropriately, it has the ability to greatly reduce transportation

**FIGURE 8.4**
Before cardboard removal.

**FIGURE 8.5**
After cardboard removal.

costs associated with garbage. This is true not only of the hauling fees/ tipping costs associated with garbage removal, but also in terms of the labor associated with the internal movement of garbage. Compactors are used to reduce the transportation waste associated with garbage. Large-scale compactors typically reduce the number of dumpster pulls from an *external* waste handler, whereas small-scale compactors typically reduce *internal* waste as employees accumulate, collect, and transport waste.

## SMALL-SCALE COMPACTORS

Small-scale compactors vary from residential appliances that compact any kind of household waste to specialized devices that hold and compact a particular waste item (such as can or barrel crushers, or balers for cardboard or other waste). All of these have the advantage of minimizing the amount of space occupied by waste while it is in the facility, and their use can help slow the decomposition of any organic waste by removing the air that it can react with. In general, a small-scale compactor can be installed in any location where a regular garbage can would otherwise be set up. An evaluation of whether it would be of real benefit to replace a regular garbage container with a compactor is primarily based on the labor and transportation needed to empty the container, either because of its remote location or because it fills so frequently.

A specialized example of a small-scale compactor can be found in the BigBelly solar compactor. This compactor is the size of a freestanding public garbage can, and is designed for autonomous installation in a remote location, such as a park. Because it continuously compacts its contents, the BigBelly compactor can easily hold many times more trash than a conventional trash can, and might only need to be emptied once a week in a space where a conventional trash can would need to be accessed and emptied every day. By reducing trips to a remote location from daily to weekly, this improvement potentially saves many hours of staff time and significant vehicle usage. BigBelly includes a calculator on its website to calculate the payback period on its units from savings in labor and fuel use.[1]

## STYROFOAM DENSIFIERS

Like a cardboard baler, the Styrofoam densifier compresses waste Styrofoam into a form that is a recyclable commodity, in this case by melting it and allowing it to cool into solid blocks or simply compressing it mechanically. For industries that use significant amounts of Styrofoam cups—such as casinos—the ability to densify and sell Styrofoam represents a significant portion of the waste stream that can be converted directly into a sellable commodity, as well as clearing out volume in their waste stream. Densifiers are not prohibitively expensive, and could be profitably installed by many industries. In many cases, a densifier will have a return on investment (ROI) of less than a year. A Styrofoam densifier can condense waste Styrofoam down to 1/40th of its original volume, making it resalable by removing the air.[2]

## REDUCING DUMPSTER PULLS—EXTERNAL WASTE HANDLING COSTS

The direct cost of external waste hauling is only reduced if the lower volume of garbage results in fewer dumpster pulls. As with residential garbage collection, most business waste hauling is regularly scheduled and fixed by contract, with a set number of dumpster pulls contracted regardless of how much material is actually in the dumpster.

One important benefit of garbage auditing is that it allows businesses to understand how much of their garbage capacity they are actually using, and whether they have the right number of dumpsters in the right locations. Tracking garbage levels over time is required to measure this since a snapshot from a single garbage audit does not provide data on trends or long-term usage. For organizations that have regularly scheduled waste pickup (rather than pickup based on full containers), this measurement can be used to develop a picture of how much excess garbage capacity is being paid for, letting businesses optimize their waste hauling contracts so that dumpsters are mostly full but not overflowing when they are pulled. Like the savings from trash compactors, this type of optimization avoids paying external waste handlers to dump air.

In projects setting out specifically to reduce the number of dumpster pulls, a period of study is required to measure the amount of empty space left in each dumpster when it is pulled. A good method of measuring this quickly is to paint lines on the inside of the dumpsters indicating their percent full, and use these to quickly assess the amount of empty space in the dumpster at each measurement.

The best place to perform this kind of assessment is in any location where there is a bank of multiple dumpsters set side by side. Where there are individual dumpsters at multiple locations, there are potentially other advantages to consider from having easier access with less transportation. In a bank of dumpsters:

1. Measure the amount of material in each dumpster for several weeks immediately before each pickup. Record the average amount of material in the dumpsters over the whole time.
2. Multiply the total number of dumpsters by the average percent full, and multiply this number by the total number of pulls. This provides the actual number of full dumpsters' worth of material that the facility generated over the observation period (round the resulting number up to the next whole number).
3. Multiply the total number of dumpsters by the number of pickups during the observation period. This provides the number of dumpster pulls the organization actually paid for during that period.
4. Evaluate whether:
   a. The same number of dumpsters could be utilized with a less frequent pickup schedule (potentially switching from weekly to bimonthly collection)
   b. A smaller number of dumpsters could be picked up on the existing waste hauling schedule

Take into account not just the average amount of waste material, but also the single highest weekly amount recorded. The highest observed amount is really the limiting factor since the dumpster capacity has to be scaled to this (Figure 8.6).

**FIGURE 8.6**
Data from three weeks of dumpster volume monitoring.

Given this example data set:

45% full on average × 5 dumpsters × 3 pulls = 6.75 dumpsters' worth of material (round up to 7 total)

5 dumpsters × 3 pulls = 15 dumpsters' worth of capacity

Highest single week of production = 51% or 2.55 full dumpsters (round up to 3)

Based on these numbers, the facility could reduce the number of dumpsters at this location from five to three and still have plenty of capacity on the same waste hauling schedule. However, a facility would need 50% or greater free capacity in order to switch from a weekly to a bimonthly pickup schedule using the same number of dumpsters without running the risk of overflowing.

For businesses that have containers pulled only when they are full, this entire problem is not an issue. If available, this kind of on-demand waste hauling service is superior, because it allows businesses to maximize their waste hauling transportation savings and never dump a partially full container. This could be thought of as just in time waste hauling.

## WATER IN THE GARBAGE

Along with issues of air in the garbage, the presence of water in some garbage streams can be a problem as well. In solid waste streams, water's *volume* is usually not the problem. Instead, solid waste containers may not be designed to contain liquids, and the problem lies in the liquid's ability to either react with otherwise inert chemicals in the trash or spill out in transit and create a mess or hazard. In garbage audits, participants are often surprised that a significant amount of the garbage's weight can come from the liquid in partially full drink bottles, or that a major mess can be created if water from an opened drink or melted ice is present.

Dumpsters designed to be placed outside allow liquid to drain out automatically, but this is not possible for most indoor garbage cans. Because of this, the best solution for minimizing the amount of water in the garbage is usually to prevent it from entering the garbage in the first place.

## DEWATERING SLURRY

Whereas one of the principal components of solid waste by volume is air, a significant amount of the volume of liquid waste is similarly composed of water. *Slurry* and *sludge* are terms for waste products that are liquid or semiliquid enough to flow, usually a material held in suspension in water. In situations where waste is produced in the form of a slurry or sludge, dewatering the waste as much as possible before disposal lowers the cost of disposal. Depending on the composition and volume of the slurry, the simplest dewatering option is often evaporation. Settling tanks allow particulate matter to settle out of waste, allowing some portion of the water to clarify and be siphoned off. Many facilities that generate liquid wastes also house small water treatment plants that include settling tanks.

A sludge press can be a faster and very effective method of removing most of the water from waste sludge. This compresses waste sludge through a set of filters, and has the potential to remove more than 90% of the water content, drastically reducing the weight, volume, and reactivity of the waste product. This option works well for businesses that produce relatively small volumes of waste (100–200 gallons per week).

### Dewatering Slurry Example—AAR Aircraft

Due to the capabilities of the municipal wastewater treatment plant that received its wastewater, AAR's waste slurry could not be processed and was sent to the landfill. In order to reduce this waste, AAR installed a sludge press (Siemens J-Press; Figure 8.7) as an addition to its existing wastewater

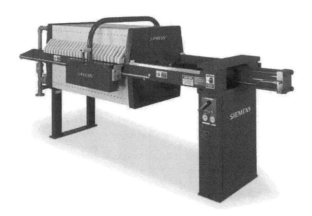

**FIGURE 8.7**
Siemens J-Press.

treatment plant. This project had a one-time cost of $20,000 for the installation and produced an ongoing annual savings at AAR of $35,000 from not landfilling waste sludge. Through the use of this new process, roughly 54,000 pounds of sludge water was diverted from the landfill annually and treated through the on-site water treatment facility. Ninety-five percent of the water was removed from the waste stream, and the relatively inert, dry waste that remained was disposed of conventionally as nonhazardous waste. AAR is also researching possible uses for the dry waste cakes, such as for oil soak. Additionally, an estimated $900 (300 gallons) of fuel is being saved annually by the hauling company for not hauling this waste.

The principal factors that unify all of these discussions around transportation waste are

- Reduce the physical volume of the trash as much as possible.
- Remove any inert filler that is in the garbage but is not really garbage (usually air and water) at the earliest possible stage, dumping only full containers of real garbage.
- Transport garbage the shortest distance using the least possible amount of labor.

## ENDNOTES

1. Learn more about BigBelly solar compactors at its website: http://bigbellysolar.com/.
2. See http://styrofoam-densifier.com/technology.htm.

# 9

## Reuse and Repurposing

There are two major approaches to reuse programs:

1. Redesigning an existing process to stop using a disposable product and using a reusable one instead (material waste)
2. Designing a new use for a waste product that had been previously discarded (garbage)

In either formulation, reuse programs usually require some level of process reengineering.

Although recycling and reuse are discussed as discrete categories, there is really a spectrum of activities that run from pure recycling to pure reuse. In between the extremes, there is a gray area where materials undergo some amount of transformation in the process of being reused (Figure 9.1).

## REUSABLE CONTAINERS

Most waste streams are primarily full of containers and packaging, and therefore reuse programs designed to phase out existing disposable products often focus first on disposable containers.

In an ideal container reuse program, the reusable container should be able to interface with the system in such a way that it does not cause any changes in operation. This avoids the need for retooling or retraining. One of the simplest examples of this is the replacement of Styrofoam drinking cups with reusable cups. While behavioral changes may be necessary on the user's part to motivate returning the reusable cup for refill, the actual interface between the cup and the drink machine is exactly the same with both kinds of cups because both cups have the same functional design.

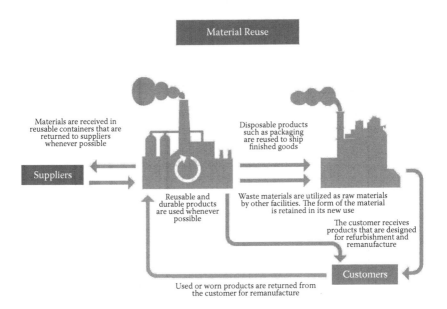

**FIGURE 9.1**

The spectrum of material reuse possibilities. Reuse opportunities exist along any of these arrows.

In situations like this, the cost of the reusable container program only involves purchasing the containers themselves, since nothing about the equipment itself has to be reengineered to accommodate the new product. This is ideal in cases where the equipment is still serviceable. If older equipment is being phased out anyway, more options are opened up for new container designs that may be more efficient.

In many scenarios involving container reuse, a new process may be necessary to clean or prepare the old containers before they can be reused. For internal processes, this labor burden falls on the organization using the containers; otherwise, it falls on the supplier or customer who returns the container ready for service. Some types of used containers will simply be empty, while others will be contaminated with whatever material had been inside.

The value of labor time vs. material costs should be calculated when considering the implementation of reuse programs. In order to justify instituting a reusable container system, the total cost of labor associated with handling and preparing the container for reuse should not be higher than the cost of the material itself. This can be a challenge since many disposable containers are designed to be extremely cheap.

As an example of a less than ideal reusable container scenario, reusable shopping bags present a situation where the reusable product does not have exactly the same form as the disposable one it replaces. Unlike the disposable plastic bags, the reusable bags are nonstandardized, and most cash registers do not have a convenient hook or clip system for the checker to easily hang the bag and fill it. Instead, these bags are usually draped inefficiently over the rack system that is designed to hold packets of disposable plastic bags. This makes the reusable container significantly less convenient for the cashier than using the regular plastic bags and impacts the speed of the process by creating extra handling time for the products.

For a reusable container system to be effective, it needs to present minimal inconvenience to the operators and, if possible, should make the operator's job easier or quicker. Reusable containers can make an operator's job easier by conveniently stacking or collapsing when empty, or by holding a material in such a way that it is simpler and quicker to pack or unpack the correct amount. For example, if the container not only holds a product, but is also designed to measure a specific amount of the product it holds, it can have the additional benefit of acting as a process control feature.

### Greenhouse Pot Reuse Project—Sam Noble Foundation

The Sam Noble Foundation is an agricultural research facility in southern Oklahoma with extensive greenhouses where researchers from around the world study plant growth and disease under very controlled conditions. Because many experiments deal with pathogens or genetically modified plants, the standard practice of the facility's staff has always been to dispose of all pots at the end of a research project, autoclaving pots and soil to sterilize them before landfilling.

However, upon further study, the greenhouse staff determined that the bulk of the pots being thrown away could simply be washed out and reused because the experiments they were used for did not represent any kind of contamination risk at this stage. In order to facilitate this, the team used available materials to build a washing/drying system for the pots in an empty part of the facility (Figures 9.2 and 9.3). Pots and materials with any pathogenic contamination concerns are clearly marked and disposed of through autoclaving. As part of the research for this project, the team benchmarked best practices in pot reuse from similar research facilities, and found that other organizations had been successfully implementing pot reuse systems without compromising their experiments. In fact, many of the researchers coming to work at Sam Noble were surprised not to be working with reused pots.

The initial goal of this project was to wash and reuse 80% of the pots and trays. In implementation, they found that they were initially able to reuse an estimated 30–40% of their pots and trays, and continue to work on expanding the project. Even at 40% pot recovery, this effort should produce a material savings of over $10,000 annually while using minimal labor (Figures 9.2 and 9.3).

**FIGURE 9.2**
Pots soaking with agitators to dislodge dirt.

**FIGURE 9.3**
Recovered pots ready for reuse at Sam Noble.

## RELATIONSHIP WITH THE SUPPLIER

Reusable container systems in particular are best used in places where customers and suppliers have ongoing relationships and empty containers can cycle back naturally to the supplier for refilling. The most ideal scenario involves a regular supply run between facilities, where a delivery truck goes back to the supplier's facility after it is emptied, and can carry back empty containers to be refilled. In Lean facilities, reusable containers can be specialized and integrated into kitting processes, particularly in situations where the supplier kits the material. In some cases, this kind of relationship allows for the development of reusable containers and packaging that better protect the product being shipped, since they are designed specifically for that product and are expected to last.

The reusable cup example demonstrates some of the supplier/customer problems inherent in reusable container systems, because it requires work and behavioral changes on the part of the customer to bring the cup back. The customer only sees a real motivation to make use of the reusable container when the cost benefits of not buying the disposable containers are passed back to him or her. In the case of reusable cups, this behavior is often artificially incentivized through a discount for bringing the cup in, or an extra charge to buy a disposable cup. In larger industrial applications, the supplier is responsible for purchasing shipping containers, so it will benefit directly from any savings in purchasing reductions.

As a side benefit, the development of reusable container systems encourages long-term work with local suppliers. Keeping the supply chain local in this way tends to produce deeper relationships between customers and suppliers, with local suppliers able to be more responsive to flexible customer needs. Purchasing from local suppliers also strengthens the local economy.

## REPURPOSING AND REUSING CONTAINERS

Cardboard is one of the most commonly used materials for disposable containers. Facilities that produce significant amounts of cardboard waste often invest in a cardboard baler to compress that waste into a product that they can sell directly as a commodity. Balers are relatively inexpensive and, for facilities with even moderate cardboard use, usually pay for themselves quickly. Balers can also be leased, with little up-front cost. Beyond just recycling, cardboard is a material that is usually reusable unless it has been contaminated. All of these cardboard reuse examples eliminate the need to purchase materials:

- **Directly reuse boxes.** Work with your suppliers to receive incoming raw materials in boxes that will also serve your business as outbound shipping boxes. This is one of the simplest reuse solutions if a reusable container exchange system with suppliers is not practical. Henry Ford famously set up a relationship like this with his suppliers, specifying the exact dimensions and construction of shipping pallets so that he could reuse the timbers as floorboards in his cars.[1]
- **Shred cardboard as packing material.** Shredded cardboard works as a packing material in virtually every situation where Styrofoam peanuts are used. Using it in this manner eliminates the need to purchasing packaging, and ships out an entirely recyclable product.
- **Build customized boxes from waste cardboard.** Some companies use waste cardboard to construct their own packaging on site. The company Soap Hope (http://www.soaphope.com) provides an excellent example of this practice, fabricating shipping packaging for its bottles of soap. It estimates the cost of its "ugly boxes" to be 50% less than purchasing new boxes for its products, labor costs included (Figure 9.4). The curve of the box fits the shape of its bottle, and is torn on the right side from being opened. The label reads: "IS THIS BOX UGLY? We made this box ourselves out of cardboard scrap from our shop floor, including shipping boxes from our suppliers and local retailers. This ugly box saves you money by reducing packaging

**FIGURE 9.4**
Soap Hope ugly box made out of waste cardboard.

costs, and it reduces Soap Hope's impact on the environment. Please recycle the materials in this package. Thanks for helping us do our part toward responsible and sustainable business practices!"

Visible cardboard reuse programs such as these can also be spun into public relations efforts as demonstrations of corporate commitment to environmental health, beyond the economic benefits of conservation. Or—as in the Soap Hope box—they can also become teaching moments to educate their customers on the benefits of conservation. For segments of the population that value conservation, this kind of detailed dedication to conservation will reinforce loyalty among their customers.

Since cardboard comes from so many sources, a combination of several approaches may be needed to fully address this waste. In very complex purchasing environments, some of the techniques discussed in Chapter 11 about tracking and reducing paper use may also be applicable to cardboard.

## SPECIALTY MARKETS FOR MATERIALS

Some waste products have special niche markets among crafters, do-it-yourselfers, or even collectors. Probably the best first step in assessing whether this is an appropriate solution for your waste is to look and see whether others are selling it on sites such as eBay, and then judge whether the price it is being sold for is adequate to justify the labor involved in selling it in this manner from a business perspective.

A great example of a waste product this applies to is metal bottle caps. Most bars and restaurants simply discard these, even if they recycle glass bottles. However, there is apparently a thriving trade in beer and soda caps on eBay for collectors and crafters. Even large, unsorted lots of used common American beer caps sell for $0.02–$0.05 per cap on eBay, and specialized or themed lots are priced much higher per cap. This means that even common bottle caps sold for reuse have a higher value per unit than aluminum cans sold for recycling.

Specialized upcycling programs like this are more labor-intensive than pure recycling programs. More care and handling may have to be taken with the materials, since their form is of value to the customer, not just their material. But, material salvaged for reuse and upcycling will also often have a higher per unit value, for exactly the same reason.

## MATERIAL EXCHANGE PROGRAMS

Containers are a good place to begin considerations of material reuse, because they form such a significant component of most waste streams, but material reuse efforts can potentially recapture the bulk of other waste through material exchange programs.

It is important to qualify the standard definition of garbage presented at the beginning of this book by saying that what is really meant by saying material is "worthless" is that it is considered worthless *in its current location*. Garbage from one process may well be a useful raw material for another process, as discussed earlier. The challenges are in making people aware that the waste material exists and is available, and in finding a cost-effective way to get that waste material into the hands of people who value it.

Material exchange programs facilitate this transfer, exchanging unneeded or waste materials between different offices, departments, or even businesses. Many institutions already have an informal internal material exchange program in the form of a shared closet of supplies, or a designated place where any surplus items may be placed to be taken by others. Even more informal material exchange programs may simply center around a single employee who everyone in the company knows as the go-to person with questions on how to rehome unwanted items. These informal systems can be effective at a very small scale, but lack the capacity to effectively exchange materials over large organizations. Since informal systems like this are also dependent on the efforts of specific people—rather than being the responsibility of specific positions—they are prone to disruption any time the personality driving them is unavailable or moves to a new position.

In order to make a material exchange program effective, company policy should specify exactly what materials need to be offered through the exchange before disposal. If materials in the material exchange program are to be available for customers outside of the company, rather than just to other departments internally, the listings of some materials may also include a price. It may be possible to sell some materials, while for others it may be profitable simply to give them away rather than paying to dispose of them. Some waste materials may also potentially be best sold back to suppliers, who can be best positioned to reprocess them into their products.

As an example of such a policy, the State of Oklahoma has specific statutes on the disposition of waste material of potential value generated by any state agency, emphasizing material exchange as the preferable way to dispose of unwanted property, but also laying out specific disposition practices for each general type of waste that is not exchanged:

### OAC 580:65-5-1. METHODS OF DISPOSAL OF SURPLUS PROPERTY

(b) Methods of disposal of surplus property are

    (1) Transfer to a state agency or authorized entity. Property is transferred to another state agency or authorized entity with or without charge as mutually agreed by both parties.

    (2) Sealed bid. Property is sold by sealed bid in accordance with 74 O.S., ß85.1 et seq. and Subchapter 15 of this chapter by the state agency.

    (3) Public auction.

        (A) State Agency. Property is sold at a public auction coordinated by the state agency, if authorized by statute.

        (B) Surplus Property Program. Property is sold at a public auction coordinated by a surplus property agent in accordance with the rules of this Chapter.

    (4) Scrap metal. Property is sold as scrap metal.

    (5) Trade-in. Property is exchanged as trade-in for replacement equipment by a state agency.

    (6) Transfer to Department for disposal by a surplus property agent. Agent disposes of surplus property, with the exception of vehicles and equipment, by any method listed in this subsection with no remuneration to the state agency.

    (7) Disposal by other means. Property is disposed of by other specified means which are deemed to be in the best interest of the state at the request of the state agency.[2]

Because the state's activities are so broad, more specific statutes detailing every individual material would be impractical; however, for most businesses, the range of materials in the waste stream is much narrower, and specific materials and disposition methods could be reasonably specified in policies.

An effective material exchange program should include a database of information on the material to be rehomed, including:

- Material type
- Disposal date—Latest date material can be held until before disposal
- Frequency—How often or at what rate will this waste be produced? Can recurring waste be aligned with a recurring customer?

- Quantity/volume/weight
- Detailed description, including condition
- Pictures, if applicable, especially for appliances or tools
- Contact person
- Location

Material exchange programs can grow internally into larger inventory control systems, where the management of both new and surplus inventories is considered to be two parts of the same problem. In some larger organizations, such as government entities, the material exchange function turns into a full-fledged surplus or property control department, where all unwanted items that are still recognized as having value are sent to be sold if possible.

The Zero Waste Network runs one of the most vigorous industrial-scale material exchange programs nationally, set up as an online forum where companies can post information about their location and waste products (http://www.zerowastenetwork.org/renewdev/). Rather than reinvent an internal material exchange program, any company looking for the public disposition of waste materials has the option to post its waste through this resource.

One of the most interesting features of large-scale material exchange programs is the fact that they create models for human industrial production that begin to look more like natural ecosystems than traditional linear industry, with consumers adapting to fill every resource niche in the waste stream. As in nature, waste products from one process are seen as a resource for another process, and an interdependent network of niche relationships evolves. Waste products are better thought of more neutrally as by-products. This is explored in more depth in the later discussion of biomimicry (Chapter 10).

## VIRTUALLY ANY WASTE PRODUCT CAN BE REUSED FOR SOMETHING

Consider human hair. As bizarre as it may seem, human hair clippings actually have a number of reuse applications in different fields.

Since hair is mostly nitrogen, and decomposes slowly, it makes an excellent soil additive as a slow-release fertilizer. Beyond the slightly ghoulish aspects of this practice, the only real concern regarding the use of human hair as a soil additive is that some hair treatments use chemicals that bind to the hair itself and would have the potential to make their way back into the food supply (this begs the question of whether they were *ever* healthy to apply to our bodies in the first place). Human hair mats are used agriculturally as mulch, covering soil and preventing weeds. These are typically made from hair imported from countries whose citizens do not tend to chemically treat their hair. Mats of human hair are also used very effectively as part of oil spill cleanup efforts since they absorb oil so well.

Hair mixed in the soil has the additional benefit of being a binding agent that prevents erosion. Similarly, "barber hair" has long been used as a binding agent in plaster, giving plaster both greater strength and the ability to flex slightly without cracking.

Beyond this, human hair also has absorptive and insulating properties.

In the early 20th century, horsehair was commonly used as an insulating material in buildings, for applications such as wrapping hot water pipes. Human hair building insulation would likely have a surprisingly high r-value, considering how warm it can be when woven into sweaters.

A good place to start researching uses for any waste product is to simply Google "uses for ___", and see what other people are already doing. Someone out there has probably already figured out how to reuse or reprocess whatever material you have, and might even be interested in receiving your waste.

## ENDNOTES

1. See http://www.greenbuildingadvisor.com/blogs/dept/green-building-curmudgeon/why-do-we-have-waste.
2. Added at 20 Oklahoma Regulation 546, effective January 3, 2003 (emergency); added at 20 Oklahoma Regulation 2547, effective July 11, 2003; amended at 21 Oklahoma Regulation, effective July 11, 2004.

# 10

## Waste Prevention through Design

Design issues are at the root of most waste, and accessing this aspect of waste prevention often involves process reengineering that goes beyond the adjustments to flow that are usually key to Lean improvements. The purpose of collecting process data in Lean work is partially to calculate the cost-effectiveness of this reengineering. Ideally, process redesign is an opportunity to simplify processes, or to eliminate some aspect of the process that causes waste. Many different approaches may be available.

### SOURCE REDUCTION EFFORTS

Much of the work in this book is focused on a discipline that is known outside of Lean as source reduction, which looks at the material flow of processes specifically to reduce and simplify inputs. One reason to pursue source reduction is to phase out materials that will only be disposed of later anyway. However, another important aspect of source reduction is the need to make direct comparisons between fundamentally different products that have the capacity to fill the same function, such as the perennial question of whether paper towels or hand dryers are better overall. In this kind of problem, the difficulty in comparing the two products originates because the cost profiles look very different: electric hand dryers have a much higher installation cost and a lower operating cost based on energy use, while paper towels have only minimal installation costs but have a relatively high material cost to use. The following source reduction worksheet provides a framework to directly compare the true costs of two products, taking into account every major aspect of their application while considering their whole life cycles: manufacture, use, and disposal.

## Source Reduction Comparison Worksheet[1]

| Factor | Product A | Product B |
|---|---|---|
| **Cost:** What is the purchase cost of each product? This includes any installation costs. This number should reflect the true costs needed to place a working unit/product on the floor. | | |
| **Warrantied life:** What is the warrantied life of the product? | | |
| **Durability:** What is the estimated life of the product in your application? This information may come from the manufacturer, maintenance records, or consumer publications. | Estimated life: | Estimated life: |
| Is the product upgradable for a longer life? | N/A<br>Yes<br>No<br>Somewhat | N/A<br>Yes<br>No<br>Somewhat |
| **Repairability:** Is it cost-effective to have the product refilled, remanufactured, or repaired? | N/A<br>Yes<br>No<br>Somewhat | N/A<br>Yes<br>No<br>Somewhat |
| Does the product have parts that are interchangeable with other models currently in use? | N/A<br>Yes<br>No<br>Somewhat | N/A<br>Yes<br>No<br>Somewhat |
| **Quantity per year:** Based on expected product life, what is the number of items needed for one year? | | |
| **Cost per year:** Cost of one unit × number of units needed in operation/durability in years.<br>*Note: Cost per year may be less than the price of one unit when the longevity of the product is greater than one year. For example, if product life is four years, then 25% of product life is used in one year.*<br>Extra material would include supporting material such as lubricants. | Cost per year of product only:<br><br>Labor cost[a]:<br><br>Resource cost[a]:<br><br>Extra material cost[a]:<br><br>Disposal cost[a]:<br><br>Total cost: | Cost per year of product only:<br><br>Labor cost[a]:<br><br>Resource cost[a]:<br><br>Extra material cost[a]:<br><br>Disposal cost[a]:<br><br>Total cost: |
| **Weight:** What is the disposal weight of one unit of the product, including packaging?<br>*Note: Absorbent products often increase in weight after use.* | | |

| Factor | Product A | Product B |
|---|---|---|
| **Weight per year:** Quantity per year × weight for the use of one unit of the product. | | |
| **Volume:** What is the disposal volume of one unit of the product, including packaging? | | |
| **Volume per year:** Number of products per year × disposal volume of one unit of the product. | | |
| **Disposal costs per year:** What are the costs for disposal of the product? Cubic yards or gallons of waste × cost for one = cost. Add under cost per year. | | |
| **Toxicity:** What is the comparative toxicity of the product in use and disposal? | N/A Low Medium High | N/A Low Medium High |
| **Worker safety:** Including servicing and repair, what is the comparative impact to worker health or safety involved in the use and disposal of the product? | N/A Low Medium High | N/A Low Medium High |
| **Labor:** What is the comparative labor expense of using the product? If labor costs and time required for each activity are known, actual costs can be estimated. Add under cost per year. | Ordering: N/A Low Medium High<br><br>Stocking: N/A Low Medium High<br><br>Servicing: N/A Low Medium High<br><br>Disposal: N/A Low Medium High | Ordering: N/A Low Medium High<br><br>Stocking: N/A Low Medium High<br><br>Servicing: N/A Low Medium High<br><br>Disposal: N/A Low Medium High |

*Continued*

| Factor | Product A | Product B |
|---|---|---|
| **Other costs:** What is the comparative resource use required through the use of the product? Quantify where possible.<br><br>Add under cost per year. | Electricity:<br>N/A<br>Low<br>Medium<br>High<br><br>Other fuels:<br>N/A<br>Low<br>Medium<br>High<br><br>Water:<br>N/A<br>Low<br>Medium<br>High<br><br>Additional materials:<br>N/A<br>Low<br>Medium<br>High | Electricity:<br>N/A<br>Low<br>Medium<br>High<br><br>Other fuels:<br>N/A<br>Low<br>Medium<br>High<br><br>Water:<br>N/A<br>Low<br>Medium<br>High<br><br>Additional materials:<br>N/A<br>Low<br>Medium<br>High |
| **Recyclability:** Is the product locally recyclable?<br><br>Is its container locally recyclable?<br><br>Does this affect costs? | Yes<br>Somewhat<br>No<br><br>Yes<br>Somewhat<br>No<br><br>Yes<br>Somewhat<br>No | Yes<br>Somewhat<br>No<br><br>Yes<br>Somewhat<br>No<br><br>Yes<br>Somewhat<br>No |
| **Recycled content:** Does the product have postconsumer recycled content?<br>   If yes, what percent? | Yes    No<br><br>____% | Yes    No<br><br>____% |

ᵃ Figures from later questions.

On an annual basis, product ___ appears to be more cost-effective and reduce waste.

### Date Code Application—Schwan's Foods

The Bon Appetit line of frozen meals is packaged in bags, and the final step of the packing process involves printing a date code on a specific area of the bag. The regulations for packaging the food require that it be bagged before the date stamp can be applied. Because of irregularities in the shape of the full bag and other problems with the process, 65% of these date stamps did not print correctly inside the box in the initial pass, resulting in extensive, chemical-intensive rework as the incorrect stamps were cleaned off manually and the bags were then fed back through the machine. A date stamp printed outside of the box was considered a defect in the process and was not allowed to ship.

This team's solution was simply to redesign the packaging so that the printer did not have to put the date stamp only in a restrictive box, but could place it anywhere on a larger, solid-colored portion of the packaging. By redesigning the packaging to drastically reduce the potential for errors, this team estimates an annual material and labor savings of at least $50,000 with no added cost of implementation. In this example, the accuracy of the printer remained unchanged, but redesigning the specifications allowed more products to pass inspection since artificial quality standards resulted in defects before.

### Foam Packaging Reduction Example—Lindsay Manufacturing

Lindsay Manufacturing makes industrial vacuum systems as well as the packaging materials to ship those vacuums. The team at Lindsay Manufacturing replaced the system that makes the shipping mold with a new design that uses roughly 20% less material. In the process of doing this, the team completely redesigned the process for making the packaging. The old packaging machine was completely manual and prone to errors and scrap, whereas the new machine is automated, making it user-friendly and time-efficient. The new mold design reduces the total volume of foam used, eliminating sections of foam that could be removed without increasing the risk of damage to the product. As a result of this project Lindsay Manufacturing is saving $10,000 a year on foam chemicals, reducing the cost of each individual mold by approximately $0.35 (Figures 10.1 and 10.2).

### New Coolant Fluid—Air Systems Components

Air Systems Components (ACS) uses saws as part of its work equipment, and these saws used a lard-based cutting fluid that produced a noxious (although technically nonhazardous) mist. This mist caused a perceived health concern for those employees working with the saws, and also created a need for extra cleaning in the workstation areas, since it settled out of the air in a gummy film. Large amounts of the old lubricant were needed since the lubricant itself did not effectively stay on the saw blade. This project utilized a new,

**FIGURE 10.1**
Old molding station.

**FIGURE 10.2**
New molding station.

much more efficient lubricant system to eliminate this mist, resulting in less cleanup time and less waste, as well as eliminating the quality of life issues caused by the mist. Beyond the cost of the lubricant, this project has produced additional savings in the cost to dispose of spent lubricant, as well as extensive labor savings from not having to clean up the oily residue and a reduction in absenteeism. ACS is currently showing an annual savings of $5,010 with an initial investment of $3,440 (Figures 10.3 and 10.4).

**FIGURE 10.3**
Original saw and cutting fluid (fluid is the liquid in plastic bottles).

**FIGURE 10.4**
**New saw and cutting fluid.**

These systems appear very similar visually, but the new cutting fluid is consumed much more slowly. Although the new fluid is more expensive by volume, it is more cost-effective since it is used so slowly.

## SIMPLIFY SUPPLIERS

An important Lean principle that is applicable in this context is that of simplifying suppliers: ordering the least total number of different parts to serve any given task, and ordering them from the smallest number of suppliers. In part, this focuses on standardizing materials as much as possible to take advantage of interchangeable parts. Look for situations where many slightly different materials are used to perform essentially the same function. While this may initially sound like it is applicable only to manufacturing, it really applies to organizations in any field that are large enough to have multiple people purchasing materials.

- Consolidating material purchasing (even office supplies) under a single purchasing contract or agreement ensures minimal costs throughout the organization.
- Standardizing equipment such as printers allows supporting materials (toner) to be interchangeable.
- Standardizing the size of a component (such as a bolt) across all product lines reduces the complexity of purchasing supplies, since the inventory of only a single type needs to be maintained.

Standardized, interchangeable parts allow for easy, cheap maintenance. Imagine, for example, if every car headlight housing conformed to a single, standard set of specifications. The cost to replace any individual headlight would be dramatically cheaper, the model needed would always be in stock, and it would be possible to recapture and reuse every unbroken headlight housing from cars that had otherwise reached the end of their lives. Nonfunctional vehicle style differences prevent this kind of durability effort from being effected, even though the shape of headlight housings has virtually no impact on the functional differences between different makes of cars.

## PURCHASING CONTROLS

Purchasing controls are important to consider, particularly in situations where expired but otherwise nondefective materials are part of the waste stream. A Lean *kanban* system can regulate purchasing and stocking to maintain a steady and appropriate level of inventory. A kanban defines when the inventory of any supply hits a reorder point, based on the rate of usage and the time required to restock.

Beyond expired materials, one other cause of waste that is possible in some manufacturing settings is unmatched batch sizes at different points in a process; if these are not aligned, the mismatch produces excess material that must be dealt with. This can result from being forced to purchase raw materials from suppliers in batch sizes that do not match up with the material needs of the process, resulting in partial containers of unused raw material that have to be either disposed of or stored.

### Sauce Loss Reduction—Schwan's Foods

Due to incompatible batch sizes with both suppliers and internal processes, Schwan's produces excessive amounts of sauces for many of its products, which then have to be used up quickly (causing overproduction), frozen (wasting energy and transportation), or disposed of (wasting materials). Realignment of these processes can be done in most cases at no cost. Projected annual savings from this project is approximately $210,000.

### Water Bottle Elimination—University of Oklahoma (OU) College of Education

This project explored eliminating bottled water from the dean's office of the OU College of Education, replacing it with a water filtration system. In 2011, the dean's office purchased 4,728 bottles of water at a direct cost of $1,270 plus labor. This department installed a large-scale Hoshizaki water filtration system capable of servicing the entire department for $4,229. Although this project has been implemented, there has been some cultural resistance to the elimination of bottled water, and some water is still being purchased. This improvement continues as an ongoing effort.

## COMPRESSED GAS LEAKS

Leaks in compressed air systems are usually considered to be energy efficiency problems, since the only waste is the energy required to repressurize the tank after the pressure bleeds away. However, in facilities that use specialized gases—such as welding facilities—leaks in gas lines are actually leaks of raw materials. Even low-pressure gas lines can have significant leaks, and a leak in a weld gas line can cause cascading problems in production, since it is released into the work environment and also presents an opportunity for atmospheric gases to infiltrate the line. A study of on-shift and off-shift weld gas usage conducted by a Lean Institute project team revealed that nearly as much gas was being lost to leaks as was being utilized by the process while in operation, with a target of 73% reduction in consumption. This also revealed attendant opportunities for quality improvement from the reduction of other gases infiltrating the weld gas line. As a general rule, all compressed air systems should be regularly checked for leaks, even if only passively to ensure that the system is not cycling during periods of work inactivity.

## PREVENTATIVE MAINTENANCE PRACTICES

Preventative maintenance is a scheduled activity, performed periodically either after a certain amount of calendar time has elapsed or after a piece of equipment has been in use for a certain number of hours or cycles. This is an essential tool to improve the performance of all types of tools, and is a key element of many Lean improvement efforts.

Replacing disposable materials with reusable ones often includes the need for maintenance practices to extend the life of those products. Although a reusable container may be durable, it will eventually wear out. Containers should be evaluated periodically with an eye for design issues and wear. The wear points on reusable containers provide information on how to eventually improve their design, especially if a pattern such as stress cracks around a handle starts to be a common problem.

Many tools and materials that were designed as single-use disposable products do not necessarily have to be disposed of at the end of their original life cycle. Maintenance practices normally reserved for durable goods can also extend the lives of disposable products. The refilling of disposable ink cartridges is a good example of this.

Research best practices for the maintenance of every type of equipment—including cleaning—and set aside an appropriate amount of scheduled time to perform that maintenance. In many situations—especially in administrative environments—equipment is replaced when it becomes old, tattered, and worn looking, even if it still functions. Even beyond preventative maintenance, respectful treatment of equipment and purchasing with durability in mind help alleviate this.

### Efficient Oil Usage—Sam Noble Foundation

Preventative maintenance standards for tractors at the Sam Noble Foundation had been set at an oil change every 200 hours of operation, as suggested by the manufacturers. However, oil analysis at 200 hours showed that the oil was still well within operational limits. This project studied the real life span of oil in these tractors by performing oil analyses on a set of tractors every 100 hours to determine when the oil actually broke down to the point that it required changing. Some of the tractors studied in this project used conventional oil and some used synthetic oil in an effort to determine how much longer the synthetic oil would last, and whether that longer life span would recoup the higher cost of the material. Reduced oil changes will not only cut down on material costs, but will represent labor reductions as well. The data collection process for this project is expected to take approximately two years.

## BIOMIMICRY

So believe me, we are not too many people on this planet. If you take the total weight of the planet's ants on one hand and the total weight of human beings on the other, you'll see that the ants' weight is four times higher. Further, they have a much shorter life span than we have. And because they work much harder physically than we do, the calorie consumption of ants equals about 30 billion people. It is clearly not about the fact that we are too many. Ants don't produce waste. They don't need to minimize waste. They produce nutrients. Again, it is a design question.

**—Michael Braungart[2]**

Biomimicry is the modeling of human systems and tools on natural processes. At a materials level, it can take the form of design innovations such as Michael Phelps's infamous "shark skin" swimsuit, which was textured modeling the properties of shark skin. This created lower drag through the water than conventional materials and gave him what some considered to be an unfair advantage at the 2008 Olympics. The same drag-reducing design principles are now experimentally being applied to both ships and aircraft.[3] At the system level, biomimicry can reflect the reorganization of material flowing through the supply chain and waste stream based on the model of ecosystems, with materials taking on the role of nutrients exchanged between organisms. There are a number of fantastic resources available for information on research into biomimicry, most notably the research database www.asknature.org. The bulk of the research available through this site focuses on very specific engineering advances through mimicry of natural systems, with links to the original research provided.

Many of the principles discussed in this book at the intersection of Lean and sustainability are fundamentally biomimetic. Biomimicry is often tied to the engineering of materials or the design of tools; however, its application at a system level to virtually any field can be summarized in a few key principles:

- Use waste as a resource.
- Gather and use energy and resources efficiently.
- Diversify into niche markets that allow for the full utilization of an environment's resource potential.
- Network and develop cooperative arrangements with other organizations in the environment.
- Constantly evolve in response to the environment.
- Do not toxify your environment. The by-products of your activities should not prevent those activities from taking place in the future. This is at the heart of sustainable practice.

Although not precisely a part of biomimicry, biological processes can be incorporated into mechanical processes, often in the field of waste disposal through bioremediation. Bioremediation specifically utilizes specially engineered biological systems to consume the waste products that would otherwise have to be disposed of. In many fields, the line between biological and mechanical is increasingly blurry anyway.

## ENDNOTES

1. This worksheet is derived from the "Source Reduction Now" material produced by the State of Minnesota: http://www.pca.state.mn.us/index.php/view-document. html?gid=4747.
2. Braungart, M. Is sustainability boring? DIZAJN CAFE. See http://www.dizajncafe. com/en/news/1/eco-design/40. See also http://www.biomimicryinstitute.org/about-us/staff.html.
3. See http://travel.cnn.com/lufthansa-tests-shark-skin-881186.

# 11

## Paperwork Reduction

Paperwork reduction is broken out here as a separate discussion because it is really a special category of garbage and material waste, and responds well to slightly adapted versions of the general tools outlined elsewhere in this book. From a Lean perspective, the cost of paper itself accounts for only a small fraction of the total true cost of using paper in an office. The cost of using paper is often an order of magnitude more than the sticker price of the material. Consider that even at minimum wage, the cost of a worker spending three minutes handling a piece of paper is 10 times greater than the cost of the printed paper itself.[1]

From a material waste perspective, it is best to think of paper as a disposable container. It stores and transmits a commodity: information. With the exception of things like diplomas, the value of the paper is almost never in the paper itself, but rather in the information that it contains. We take what we need from the information that the paper carries, and have to dispose of the container when the information is obsolete or no longer needed.

### WHY DO WE USE PAPER?

Paper is the way we have traditionally operated administrations, and older processes are almost always built around it. In most offices paper is not only a habit, but the infrastructure around which information flow is built. One important thing to keep in mind while evaluating paper-based systems is that when the system was established, paper was the optimal method of information transfer. This can present a very significant cultural inertia to overcome. It is important to keep in mind that the paper itself is not the process; it is only the vehicle that the process uses.

Paper is used in an office to do two things: transmit information and store information about processes. In today's offices, most paperwork processes are better thought of as physical extensions of electronic processes, since paper is usually either printed from a file generated on a computer or used to collect handwritten data that are then entered into a computer. The paper is an extended computer interface, so it makes sense wherever possible to work directly with the computer rather than retranscribe information.

## ADVANTAGES AND DISADVANTAGES OF PAPER

Paper has a number of advantages in areas not yet mastered by electronic technologies for information transfer, storage, and processing. However, for many of the advantages of paper, there is a corresponding disadvantage. Keep an open mind when approaching paper, because in some cases, paper may still be the most effective method of transmitting information, especially where there is not a long-term or ongoing relationship with the person receiving the information.

### Advantages

- Paper is reliable. (Paper never crashes, and its file format never becomes obsolete.)
- Paper is traceable. (Physical handoffs can aid in accountability, and physical delivery through services such as registered mail can be used as proof that someone received a piece of information. Electronic information flow is becoming increasingly traceable as well.)
- Paper is disposable. (It is easy to print off information and give it to someone on the spot without needing to get the paper back.)
- You can write on paper by hand. (It is easy to take notes and mark up draft copies of documents.)
- Everyone is comfortable with paper.
- Paper documents can reach people who have no access to computers.

## Disadvantages

- It is time-consuming and difficult to search for information in paper files.
- Paper is heavy and bulky to store and manage.
- Paper documents can only be accessed by workers on site. (There are no remote or shared access options.)
- Archived paper tends to degrade (mold, mice, fire, water, etc.).
- Because of their bulk, archived paper documents are often stored off site and are difficult to access.
- Paper documents require physical destruction to protect data security.
- Routing paper documents adds unnecessary time and labor to processes.
- Storing paper documents tends to lead to redundancy (multiple copies in different locations).
- Physical documents and reports are costly and labor-intensive to prepare.

## ACTIVE PROCESSES VS. RECORDS RETENTION

The problem that initially draws attention to the need for paperwork reduction is usually in records retention. Degraded, unorganized, difficult to access files in a storage facility may hit the limit of storage capacity, or some crisis involving documentation may shine a light on records retention as a problem. The impetus for paperwork reduction probably centers around getting this mound of papers in storage under control, so the natural temptation is to begin by addressing these stored files—avoid this temptation. The files in storage are a symptom of a poorly designed process, not the root cause.

Think of paperwork accumulation as being analogous to a plumbing problem: if your toilet were overflowing, your first action would be to shut off the water before you started cleaning the floor. Likewise, deal first with the flow of new paper being generated by your processes, and then set policies for the retention and destruction of existing files that require minimal effort to implement. If you stop the flow of new paper into the files, and have good policies in place for the destruction of old files, the retention problem will essentially take care of itself as old files are phased out.

## ANALYZING PAPER USE WITH A PAPER AUDIT

Just as the best place to begin the analysis of garbage in general is with a garbage audit, the place to begin paperwork reduction is with a paper audit. An effective paper audit needs to be conducted over an extended period of time in order to capture the full variety of processes that make use of paper. In most office environments, a paper audit should be conducted over the course of an entire month of operation to allow monthly processes to cycle through.

To conduct a paper audit, create a spreadsheet that tracks the movement of all paper across *your personal desk* for a month. Every document you personally create or touch should be listed. This individualized approach allows multiple people in the same department—or spread across an administration—to conduct the audit independently and then compare notes. Problem processes that really bear examination will show up significantly on multiple individual paper audits.

The following model shows the basic elements to include in this spreadsheet, although details may vary in different institutions. From a Lean perspective, the most important element that a paper audit can document is not the paper itself, but rather the labor that is associated with the paper use. Document the time you spend handling the paper (including times for such tasks as walking to pick up printed documents). Focus on documenting the time that directly involves paper, rather than the time spent working with a computer to generate the information that will be printed. This will help identify only the labor issues tied to the paper itself (Figure 11.1).

The results of a paper audit can be used to prioritize efforts at paperwork reductions by identifying not only problems, but also areas where change can be effected. Processes where two or more of the following factors converge are natural starting places for improvement:

1. Relatively high touch time.
2. Relatively high paper volume.
3. Relatively high number of repetitions.
4. The process is locally controlled rather than remotely administered.

Paperwork improvement projects should be treated just like any other process improvement project once a problem is identified.

Although there are many types of process maps, one of the most effective to use with paperwork is the responsibility or swim lane map. The

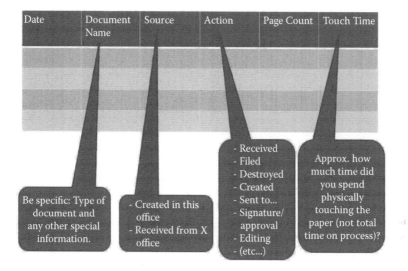

| Date | Document Name | Source | Action | Page Count | Touch Time |
|---|---|---|---|---|---|
| Be specific: Type of document and any other special information. | - Created in this office<br>- Received from X office | - Received<br>- Filed<br>- Destroyed<br>- Created<br>- Sent to...<br>- Signature/approval<br>- Editing<br>- (etc...) | Approx. how much time did you spend physically touching the paper (not total time on process)? |

**FIGURE 11.1**

Spreadsheet model for paper analysis. This could be used to track the paper crossing each individual desk in a department.

map itself looks like a swimming pool, with each department being represented by its own column or lane. This type of map is especially useful in the context of paperwork processes because it highlights points where handoffs occur between different departments.

Figure 11.2 shows an example swim lane map for an administrative process. The basic flow outlined here can be fleshed out with data, showing the touch time, wait time, amount of paper produced or retained at each step, etc.

## RECORDS RETENTION POLICIES AND DOCUMENT DESTRUCTION

Records retention policies must be in compliance with whatever regulating bodies demand documentation from the process. Beyond that basic goal, records should be retained:

- In the most cost-effective way considering materials, labor, and disposal
- In the most accessible way possible (least labor and time required to access archived information when needed)

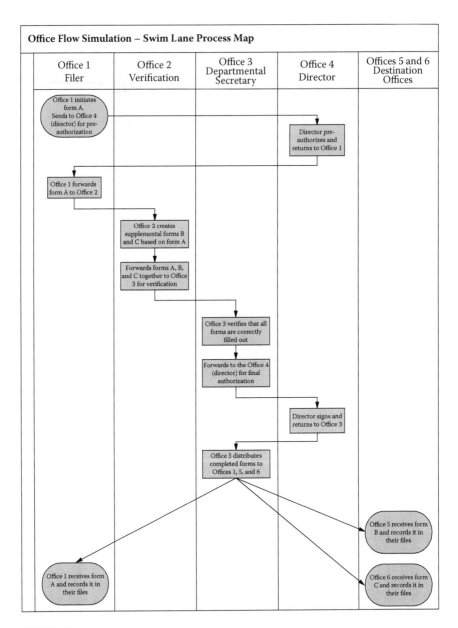

**FIGURE 11.2**

Simple swim lane map model. Each column on the map represents a different office or department in the process. This type of process map highlights handoffs between departments and can be fleshed out with data about touch time and wait time.

- With the greatest transparency
- In the most secure way (least likely to become corrupted or stolen)
- Using the smallest amount of physical or digital space possible (minimal redundancy, saving office space, and avoiding off-site storage)

Establishing a set of standards for document destruction is also a vital part of a records retention policy. For each type of document, there should be a clear retention/destruction standard that is based on a verified documentation need. For example, financial records might have to be retained for three years for tax purposes, at which point they can be destroyed. Establishing these standards at the beginning of paper reduction efforts will minimize the labor of going through the old files, since this allows many of the hard copy files to be phased out through scheduled destruction rather than being scanned and retained for a short period. As a general rule of thumb, do not scan any existing document that has to be retained for less than a year. Once a records retention policy is put in place, nothing should be scanned into storage without a clearly marked destruction date, including "do not destroy" if that is really accurate. In general, it is best to destroy documents as soon as possible based on retention requirements. For auditable processes, audit findings can come from any retained information, even if it could have legally been purged at some prior point. Destroying documents at the earliest possible dates reduces this risk.

Any document that does need to be retained should be purged of unnecessary information prior to storage. Sensitive personal data should be redacted from documents in any situation where those data are no longer explicitly needed for reporting. These data are especially important to remove before scanning, since scanned documents will be much more accessible than paper files in storage. Prior to the year 2000, social security numbers were commonly used as personal identifiers, so any personnel documents older than that date will likely include this and need to be treated with extreme care. Along with redaction, files should also be purged of unnecessary papers prior to scanning, simply to minimize the size of stored files and aid the accessibility of important information later.

Keep in mind that the records retention schedule applies to scanned electronic records as well as paper ones, and materials in the electronic file system should be marked with destruction dates too. Outlook can be a good tool for managing this.

## BUILD A ROBUST ELECTRONIC FILE SYSTEM

There is a common complaint that once an office "goes paperless," it may instead tend to use more paper than it did before. This is a real phenomenon, and is typically driven by a lack of trust in the effectiveness of the electronic filing system. Creating a robust electronic filing system is an important part of making a paperless office function, and a well-designed electronic filing system can be much easier to navigate and find materials in than a paper system. Many high-speed scanners come with a software package that makes it simple to scan a document to a specific location, rather than having to laboriously scan, name, and then file it electronically afterwards. However, a robust electronic filing system can also be created manually on a shared network drive without using any special management software.

As its basic factors, a robust electronic filing system needs the following:

- A set of standardized file name conventions, so that everyone using the system will always be able to find files
- An accurate map of the folders on the network drive, including a list of where each kind of document belongs
- A method to hold people accountable for properly filing, as with paper files (this can include audits)

## COMMON SOURCES OF OFFICE WASTE RESULTING FROM PAPER USE

The following list highlights a few of the most common office processes that use paper ineffectively, and are typically holdovers of outdated information technologies. If any of these problems occur in your office, working on them would provide a good starting point to minimizing the impact of paper.

**Fax machines:** No office with a computer and scanner should design new processes around paper faxes. Faxing is most useful when the population served by the process may not have computer access, and while it should be an option, it should not be the primary way of receiving information. In offices that still rely on faxes, eFax software allows fax functionality from a computer without ever making the faxes paper. These electronic fax services typically have monthly subscription fees and are good for fax-intensive processes.

**Paper forms:** If an office uses paper forms that people fill out and return, they can be set up as "fillable form PDFs," made available online, and accepted back via email. This does not mean that the paper form need no longer be accepted, but only that people who are submitting the form will have an electronic option. The availability of electronic forms also makes an office much more accessible to its customers (forms will be available at a distance and after hours).

**Wet signatures:** Legally binding electronic signatures are now accepted in many places that previously only accepted wet signatures. The need to sign documents by hand drives a great deal of printing, much of which is immediately scanned and destroyed again afterwards. This is a de facto electronic signature anyway. Everyone with signature authority should be set up with something comparable to Adobe's digital signature system, meaning that those PDF forms will not have to be printed and scanned just to be signed.

**Snail mail lists:** If your office mails hard copy materials to a list:

- Verify that all addresses are valid (purge bad addresses from the list). The post office offers this service for commercial mailing lists at no cost.
- Verify that multiple recipients do not live at the same address, or that there are not multiple versions of the same person on the list with minor formatting variations.
- Consider transitioning to an electronic version of the document.
- Send a request such as a postcard with the next issue asking people on the list to "opt in" to continue receiving paper mailers (*not* "opt out" of the paper mailer). This requires active participation on the receiver's part to continue receiving hard copy mail and allows those who strongly prefer paper or have no computer access to continue receiving the information. This can be used as a tool to either phase in an electronic version of the document or simply to cull uninterested parties from the list.

**Courtesy copies:** In larger organizations with multiple departments, one department is usually the office of record for any specific kind of document. If your office is not the office of record, you have a courtesy copy of no official value. Many paper processes commonly generate courtesy copies of documents at each office the document moves through, leaving a physical record that the document passed through those hands. This can instead be done electronically. Many offices make "tickler files" of documents in process, creating a temporary copy of an unfinished piece of paperwork as a reminder that the finished one is expected back. Ticklers can be created electronically more effectively using Outlook reminders, or are rendered unnecessary by some electronic document management systems.

**Excel formulae:** Internal forms and documents that involve calculations, such as travel claims, can be based in Excel, allowing the electronic document itself to perform the calculations automatically, or automatically reject data in the wrong format. These avoid not only the initial risk of errors, but also the wasted time spent in verification as other people in the process recalculate numbers to avoid sending on potential mistakes.

**Best practices for all remaining paper use:** For all of the paper use that remains, best practices of paper use can be set as the standard:

- Duplex (double-sided) printing should be set as the default print setting at all workstations.
- Black and white printing costs about seven times less than color printing per page. Set B/W as default ("draft quality" printing uses even less toner).
- Examine the format of departmental forms that must be printed. Can a form be consolidated onto fewer pages?
- Reuse or recycle the waste paper you do generate.

### Bad Mail Reduction—Tinker Federal Credit Union

At Tinker Federal Credit Union (TFCU), pieces of mail are sent to lists of members, and some of the addresses are rejected as bad. These pieces of bad mail are returned and have to be processed manually. This project's goal was to make this mailing process more efficient, by tapping into the USPS database, which gives the most current information. These bad addresses can then be flagged, stopping the mail before it goes out rather than processing it after it comes back. Total costs from this problem had been roughly $186,375 annually, and this project is estimated to have eliminated over 95% of these costs.

## REDUCING TONER USE

Toner costs are also an important part of the cost of using paper, and an examination of documents can reveal striking inefficiencies in graphic design. In examining departmental forms, look for large colored areas or other graphics built into forms that have no functional purpose and could be changed without necessarily changing the contents of the form.

The following example shows a form that was redesigned based on this principle. This form is used in the University of Oklahoma College of Continuing Education to set up new classes. Twenty-two percent of the original form's area was covered in red toner, with white lettering in these red fields. This form was redesigned, replacing the red fields with pale gray, and changing the lettering to black. Total estimated toner coverage in the redesigned form is down to 5%. Other than the color change, the content of the form was unchanged (Figures 11.3 and 11.4).

Based on toner costs and the estimated reduction in usage, this change reduced printing costs by $0.062 per page. In the process of filing, these forms are also copied on a black and white copier several times, so the toner savings repeats several times throughout its life.

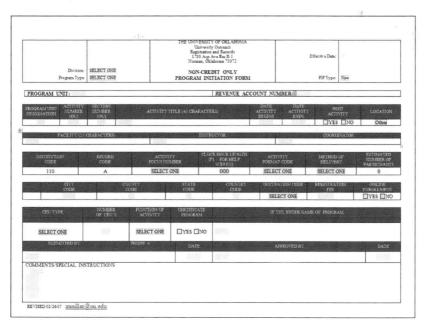

**FIGURE 11.3**
Original form with dark decorative fields.

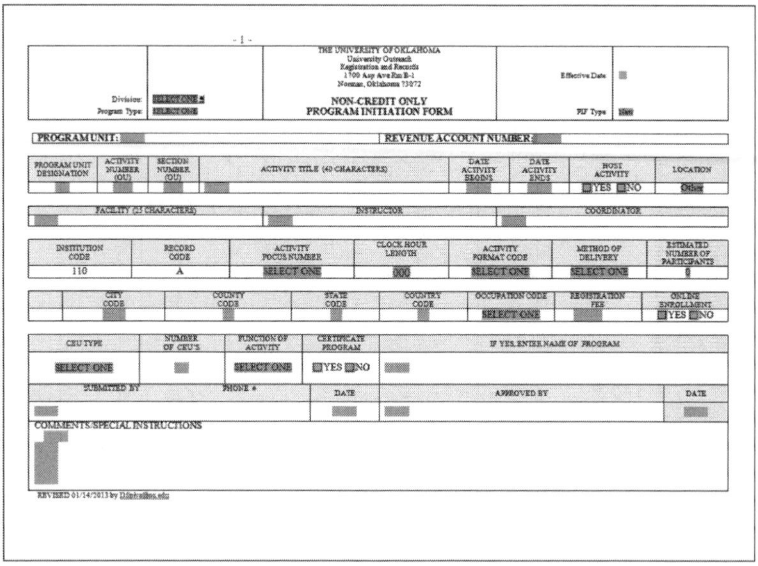

- 1 -

| | | | | |
|---|---|---|---|---|
| Division: | SELECT ONE | THE UNIVERSITY OF OKLAHOMA<br>University Outreach<br>Registration and Records<br>1700 Asp Ave Rm B-1<br>Norman, Oklahoma 73072<br>**NON-CREDIT ONLY**<br>**PROGRAM INITIATION FORM** | Effective Date | |
| Program Type: | SELECT ONE | | PIF Type | |

**PROGRAM UNIT:** | **REVENUE ACCOUNT NUMBER:**

| PROGRAM UNIT DESIGNATION | ACTIVITY NUMBER (OU) | SECTION NUMBER (OU) | ACTIVITY TITLE (40 CHARACTERS) | DATE ACTIVITY BEGINS | DATE ACTIVITY ENDS | HOST ACTIVITY | LOCATION |
|---|---|---|---|---|---|---|---|
| | | | | | | YES NO | Other |

| FACILITY (25 CHARACTERS) | INSTRUCTOR | COORDINATOR |
|---|---|---|
| | | |

| INSTITUTION CODE | RECORD CODE | ACTIVITY FOCUS NUMBER | CLOCK HOUR LENGTH | ACTIVITY FORMAT CODE | METHOD OF DELIVERY | ESTIMATED NUMBER OF PARTICIPANTS |
|---|---|---|---|---|---|---|
| 110 | A | SELECT ONE | 000 | SELECT ONE | SELECT ONE | 0 |

| | CITY CODE | COUNTY CODE | STATE CODE | COUNTRY CODE | OCCUPATION CODE | REGISTRATION FEE | ONLINE ENROLLMENT |
|---|---|---|---|---|---|---|---|
| | | | | | SELECT ONE | | YES NO |

| CEU TYPE | NUMBER OF CEU'S | FUNCTION OF ACTIVITY | CERTIFICATE PROGRAM | IF YES, ENTER NAME OF PROGRAM |
|---|---|---|---|---|
| SELECT ONE | | SELECT ONE | YES NO | |

| SUBMITTED BY / PHONE # | DATE | APPROVED BY | DATE |
|---|---|---|---|
| | | | |

COMMENTS/SPECIAL INSTRUCTIONS

REVISED 01/14/2013 by [illegible]

**FIGURE 11.4**
Improved form with reduced toner use.

Another toner reduction option involves creating "pretty print" documents, especially for websites and presentations. Pretty print options allow a document to appear in a colorful format on a screen, and print out in another format that leaves out most of the noninformative graphics. There are software print plug-ins that can perform this function on any website or document prior to printing.

**Printing Reduction Example—American Fidelity Insurance**

Salespeople at American Fidelity insurance company process applications on laptops. If successful, the application was saved electronically, but if the application failed, it printed automatically. This team implemented a process so that if the application failed, the sheets that normally would print were instead saved electronically. The jet issue process and printing project improvement are saving American Fidelity 6 million sheets of paper annually, which reflects a yearly savings of $180,000.

## PAPERWORK REDUCTION VS. PAPERLESS OFFICE

Waste paper reduction efforts often assume that a truly paperless office is the most desirable goal. However, in some situations paper is the most effective way to convey or process information. Instead of thinking primarily about the elimination of physical paperwork from our business processes, it is more useful to look at ways to streamline information flow as much as possible. In most cases, approaching information from this perspective will automatically remove paper as a by-product of the changes, because paper is usually not the most efficient method. Approach this thinking of the paperless office as involving gradually less paper rather than no paper now.

Becoming a paperless office is a transitional process that requires the redesign of many individual paper processes, sometimes dozens of small processes one at a time. As such, the work to reduce paper will take time. Many offices working toward paper reduction will set a long-term target to allow integration and acclimation of new ideas as part of a gradual culture change: a 5–10% reduction in paper use per year for the next 10 years is an attainable goal.

As with other garbage, the most important aspect of a paper reduction effort is to spend less time and energy handling materials, thus reducing the amount of paper and the work that goes along with it.

## ENDNOTE

1. Minimum wage is calculated at $7.50/hour, or $0.125/minute. Paper itself typically costs $0.02/page, plus another generous $0.02 for printing costs. Labor for three minutes would be $0.375; paper would be ~$0.04.

# 12

## Regulated Waste Segregation

Much of the focus of waste reduction is on attacking the sheer volume of waste produced, but correctly disposing of regulated waste can have equally profound impacts on the real cost of waste disposal without reducing actual volumes. Beyond sorting recyclables for reclamation, this is a factor in industries where waste segregation is required because some portion of the waste is restricted beyond the regulations applying to conventional waste.

Working on the problem of regulated waste segregation does not reduce the net volume of trash being produced by a facility, but it has the potential to reduce the financial and environmental costs associated with that waste by processing all waste only to the minimum requirements. The handling and disposal of regulated waste is usually many times more costly and energy-intensive than conventional waste disposal, so getting conventional waste out of these regulated waste streams can have benefits beyond reducing the overall volume of material.

There are three major categories of regulated waste that apply to many facilities:

- Hazardous materials
- Regulated medical waste
- Confidential paperwork

Although these three types of regulated waste are generated by different sources, the problem of proper segregation is essentially the same for all of them. Each of these types of regulated waste is more costly and energy-intensive to process than the conventional waste stream. Each of these is also more complicated as a waste stream, since specific regulations require different disposal methods for different components.

With all kinds of regulated waste, there can be two types of problems in improper disposal: contamination of the conventional waste stream with regulated material, and the extra cost of processing involved when conventional waste is disposed of in the regulated waste stream. Of the two, contaminating the conventional waste stream with regulated material is a much more serious problem, so there is a tendency to dispose of any questionable material in the regulated stream when in doubt. In some cases, an abundance of caution will lead to *all* waste being processed as regulated waste to avoid potential slips. This is most commonly seen in the disposal of paperwork. In many offices all paper waste is sent through a document shredding service to avoid any possibility of information leaks.

In all fields, these problems are driven by five major factors:

- Sorting waste is considered too time-consuming to be worth the effort.
- The risk of improperly releasing regulated material is seen as higher than the cost of overprocessing it.
- Policies for exactly what requires regulated disposal are vague or not well understood.
- Workers never see the costs of regulated waste disposal and do not necessarily know that there is any cost difference in the different waste streams.
- There is little or no accountability for improperly disposing of conventional waste in the regulated containers, because access to those containers is restricted and their contents may not be examined. These barriers protect both the materials and the employees.

The importance of each of these factors varies in different fields, being especially dependent on the number of material types that become waste. Fields with more complex waste streams will tend to see all of these problems exacerbated, increasing the likelihood of confusion over the disposition of individual items. The more complex the waste, the clearer the disposal system must be.

## RESOURCE CONSERVATION AND RECOVERY ACT

The Resource Conservation and Recovery Act (RCRA) is the primary U.S. statute governing the disposal of all kinds of waste. Enacted in 1976, this statute gives the Environmental Protection Agency (EPA) authority to regulate the handling, storage, transportation, and disposal of material waste in all its forms. So while this book has referred to a colloquial distinction between conventional and regulated waste, it would really be more appropriate to say that *all* waste disposal in the United States is regulated, with restrictions placed on potentially dangerous materials.

Beyond simply giving regulatory authority to the EPA, RCRA lays out a waste reduction ethos. The act lays out a regulatory framework in which reuse and recycling of materials are made more desirable than simple disposal. One way in which it does this is through the fact that the EPA's definition of waste purposefully excludes materials that are being reprocessed within a manufacturing process. This reduces unnecessary interference with internal material recovery operations inside companies. RCRA uses the following test to determine whether something is really a waste product as opposed to work in process being recovered[1]:

- Is the material typically discarded on an industrywide basis?
- Does the material replace a raw material when it is recycled, and how similar is its composition to the raw material?
- What is the relation of the waste recovery to the main activity of the facility?
- Is the material handled in such a way that it is secure, and release to the environment is minimized?
- What is the length of time that the material is accumulated before reprocessing?

Material that is judged to be remanufactured into a raw material by the facility is regulated through RCRA, while material that is simply recovered is not.

One of the most important provisions of RCRA is the establishment of cradle-to-grave regulation of hazardous materials, regulating the manufacture, transportation, storage, use, and disposal of hazardous materials through their entire life cycle.[2] Cradle-to-grave monitoring of hazardous materials creates accountability throughout the industries that deal with those materials.

Beyond the EPA's regulatory term, the phrase "cradle to grave" is also sometimes set up in contrast to the cradle-to-cradle design philosophy, as opposing models of manufacturing. Cradle-to-cradle essentially closes the loop of material flow through reuse and recycling. In this context, cradle to grave refers not to the EPA's standards, but to the fact that most manufacturing processes are linear, harvesting virgin raw materials and then creating products out of them that are landfilled at the end of their lives. Created by William McDonough, cradle to cradle adopts a biological model in which manufactured materials are seen as "technical nutrients," drawing parallels to the exchange of organic nutrients between biological systems in nature. As in nature—as seen earlier in the discussion of biomimicry—material waste products from processes are recaptured by other processes rather than becoming waste. Once a nutrient is part of a system, it is retained by the system until some process consumes or transforms and it no longer retains its original properties.[3]

## DEFINING HAZARDOUS WASTE

*Hazardous waste* can be loosely defined as any waste material that is prohibited from being either sent to the conventional landfill or disposed of through the sewer to municipal wastewater because of its chemical makeup. Hazardous waste also includes any chemically inert waste product that poses a physical danger to the people interacting with it.

According to the EPA, a waste product is considered hazardous if it meets one of the following criteria:

1. Listed waste: The waste appears on one of the published lists in the Code of Federal Regulations (40 CFR Part 261).
2. The waste has one or more of the following properties:
   a. It is highly flammable (known as ignitable waste). Paints, solvents, and fuel are examples of this category; wood and paper are not.
   b. It is highly corrosive (very low or very high pH).
   c. It is explosive or produces toxic fumes or vapors, especially when in contact with water or under heat or pressure.
   d. Absorption by humans is harmful or fatal, and it has the capacity to leach into the soil or groundwater.

The EPA's definition of *hazardous* includes qualifying terms such as *highly* that leave room for interpretation. If unsure whether a material qualifies under one of these categories, it can be tested directly, or the material safety data sheet (MSDS) associated with the material can be consulted for any hazardous properties.[4] The state Department of Environmental Quality should be contacted for local disposition standards of hazardous material if those standards are not already known.

Beyond these general classifications of hazardous waste, the EPA has another category for hazardous waste products that tend to be produced in small amounts by virtually all organizations. The EPA's "universal wastes" include batteries, lightbulbs containing heavy metals, waste pesticides, and mercury thermostats, with individual states optionally adding other wastes to the universal waste list. Although these materials are usually not the *primary* waste product of any organization, their disposal is regulated as a hazardous waste, since they have the capacity to leach toxins into the soil and water and tend to accumulate in landfills.[5]

Keep in mind that many of the waste chemicals branded as hazardous by the EPA are hazardous because they are highly reactive or volatile in some way. Because of this energy or chemical potential, many hazardous waste products themselves are good candidates for recycling, reclamation, or reuse.[6] Determining the recycling or reuse potential of hazardous waste is one of the best reasons to analyze the contents of a hazardous waste stream. Recycling or otherwise reusing hazardous waste is a regulated activity that the EPA strongly encourages.[7]

## MEASURING HAZARDOUS WASTE

Hazardous waste cannot be physically sorted after it has been segregated, because further contact with it is prohibited. In some cases there are also regulations over how long the waste may be kept on site before disposal. However, there are several different ways to approach measurement as extensions of the garbage audit model:

- The material can be audited as it is being produced, such as with a material log filled out by the people contributing the material.
- Separate containers can be weighed and set out for each specific type of hazardous waste material. These can then be reweighed when they are full or at the end of a specified amount of time.
- In cases where hazardous waste is delivered to a third-party waste handler for disposal, the waste handlers should be able to provide a report on the volumes and materials that they have accepted. Even if

this does not provide a detailed material breakdown, it should provide information on long-term trends and volumes. This information should also be available from company records.

If hazardous waste materials are placed in a holding area to outgas before their containers are sealed for disposal, they should be weighed both before and after the outgassing process.

## EXPLORING NONHAZARDOUS ALTERNATIVES

Similar to garbage in general, the first questions to ask of everything that appears in the hazardous waste streams are: Does this have to be *hazardous* waste? Is there a way to complete this process without using hazardous materials? These go beyond merely asking whether the material is being properly sorted as hazardous or conventional waste (e.g., making sure that no inert cardboard is placed in a container designated for oily rags). This can explore mechanical processes as alternatives to chemical ones (e.g., sandblasting instead of chemical treatment to prep parts for painting), or question whether a process is really of value to the customer (e.g., questioning whether a part *really* needs to be painted at all if a clear coat is applied on top of the paint). In other cases, there may simply be a less toxic chemical process that serves the same function, such as switching from oil-based to water-based paint. This is also an opportunity to verify whether the standards that are *believed* to be customer-driven requirements really are, especially since less hazardous processes may well provide lower-cost products.

In some cases, direct comparisons of the use of hazardous vs. nonhazardous materials can only be seen by creating a future state map for the process that explores the overall impacts of the change, since the process itself may change significantly in more ways than just material type. In other cases, two different products that serve the same function can be compared directly using the source reduction worksheet from Chapter 10. After auditing hazardous waste materials, it would be useful to go through a variation of the garbage interrogation process with each significant component of the waste stream. As in the conventional garbage interrogation process, a waste only has to be hazardous material if the answer to all of the following questions is no.

| Material: | | |
| --- | --- | --- |
| Weight/amount uncovered in audit: | | |
| Estimated annual volume: | | |
| 1. Is the disposal out of compliance with regulations or company policies for this type of material? | Yes/No | If yes, explain why. |
| 2. Is the material recyclable but not recycled? | Yes/No | If yes, what barriers are preventing recycling now? |
| 3. Is this an inert material that has only become hazardous because it has been contaminated? | Yes/No | If yes, explain the contaminating process and material. |
| 4a. Could this material have been reused or reprocessed in your facility? | Yes/No | If yes, where and how? |
| 4b. Could this material have been reused or reprocessed by a different industry to your knowledge? | Yes/No | If yes, where and how? |
| 5. Are nonhazardous alternative products available that could serve the same function within the existing process? | Yes/No | If yes, compare costs using source reduction worksheet. |
| 6. Could a mechanical or nonchemical process be used to replace the process generating this waste? | Yes/No | If yes, draw a future state value stream map. |
| 7. Especially for older processes, would an equipment upgrade allow the process to generate less hazardous waste? | Yes/No | If yes, what is the anticipated useful life of the existing equipment? |

## REGULATED MEDICAL WASTE

As with other hazardous waste, the regulations governing regulated medical waste are driven primarily by the EPA.[8] The exception to this is in the disposal of unused narcotics, which are regulated instead by the Drug Enforcement Agency. Medical waste is complicated by the fact that medical facilities deal with a whole spectrum of wastes that require different kinds of handling. These include soiled linens with a potential range of infectious properties, sharps, medications, radioactive waste, and human bodily fluids.

In the year 2000, the American Hospital Association (AHA) and the EPA met to discuss the problem of excessive amounts of medical waste in U.S. landfills. A driving force behind this discussion was the increased cost of disposing of medical waste due to extra protective measures placed on landfills. These changes were driven in an effort to prevent pathogens from entering the groundwater supply. Both the AHA and the EPA were concerned that an estimated 44% of material labeled as regulated medical waste was, in fact, processed unnecessarily and could have been disposed of in conventional waste.[9]

The EPA lists seven qualifiers for a material to specifically be regulated as medical waste, as opposed to more general hazardous waste. The bulk of infectious wastes are sterilized through autoclaving before being land-filled. Each of the following categories contains materials with a broad spectrum of chemical and physical characteristics, and the physical characteristics of the materials are usually the prime determinant in how they are collected to be processed (i.e., sharps, liquid wastes, and solid wastes)[10]:

1. **Microbiological wastes:** Infectious wastes that have the ability to cause disease in humans
2. **Human blood and blood products:** Note that this category does not include human waste itself. So a container of human urine from a patient could simply be flushed down the toilet and the container placed in the conventional waste unless there was some additional reason to classify it as infectious or chemically dangerous (e.g., chemotherapy drugs).
3. **Pathological wastes of human origin:** Human bodies and body parts.
4. **Contaminated animal wastes:** Animal bodies, body parts, associated materials believed to be infectious.
5. **Isolation wastes:** Waste products that have contacted highly infectious animals or people.
6. **Contaminated sharps:** A sharp is defined as any piece of regulated medical waste that has the potential to puncture a plastic bag and has to be transported in a sturdier container.
7. **Uncontaminated sharps.**

Materials produced outside of standard medical facilities—as in the case of home health practitioners—are also categorized as regulated medical waste.

Note that the regulated medical waste categories do not include any chemicals associated with the practice of medicine, but focus instead on pathogens. Because of their chemical components and infectious natures, many regulated medical waste products are also classified as hazardous materials in their own right. Also like other hazardous waste, regulated medical waste cannot be transported without a shipping document attesting to its contents and origin.

See Appendix C for a detailed look at regulated medical waste disposal practices in a large medical facility.

## CONFIDENTIAL PAPERWORK

The exploration of confidential paperwork should begin with finding the specific regulations that govern the storage and disposal of sensitive information within your organization. There is no objective standard for exactly what information is considered confidential, and in different organizations this may range from employee personal information to medical information, budget information, or even proprietary research and product information.

In the case of sensitive paperwork, the material itself is not hazardous, so an audit of the confidential paperwork waste stream can be carried out by any employees who have clearance to see the information it contains. This audit can be conducted as the material is produced using the spreadsheet model outlined in Chapter 11 for paperwork analysis, adding a new column showing whether or not the document is confidential. An autopsy audit can still break down the paper contents of the waste stream by different types of documents or different people/departments producing the documents. This audit should list a page count per item. Additionally, this audit should specify whether each document type found in the confidential paperwork bin really required confidential treatment, or could have been disposed of as conventional paper waste.

If personal shredders or a professional shredding service are typically used to dispose of sensitive paperwork, this practice should be discontinued during the collection period for the audit, and all paper that would have been shredded should be collected and shredded afterward.

To experienced garbage auditors and identity thieves, nothing is more obvious in the garbage than paperwork that contains sensitive

**FIGURE 12.1**
Confidential paperwork recovered from the garbage.

personal information, and a trained eye can easily spot it. As an example, Figure 12.1 shows a box of paperwork pulled from a public dumpster as part of a garbage audit, containing 10 years of a family's personal financial records, including copies of tax returns, copies of photo IDs, credit cards, vehicle registrations, bank account information, etc. Admittedly this is an extreme case, but because it was so extreme, it was also obvious in the dumpster at a glance. An identity thief would have had to do absolutely no work to steal this family's identities—including a military clearance— because every single desirable piece of information was present and collected together.

The owners of this paperwork were contacted following the audit and said that it had been thrown away accidentally by a friend while in the process of moving. This is a good demonstration of the security risk involved in simply *keeping* hard copy records with sensitive information on them. They pose a security risk simply by existing.

Although this is a worst-case scenario, even in more protected confidential waste streams there is risk in generating and releasing documents with sensitive information on them. Beyond the basic goal of properly sorting sensitive and nonsensitive documents for different methods of destruction, a prioritized elimination of sensitive printed documents automatically reduces this risk and lessens the work of disposal. Effort spent in reducing the production of sensitive documents is more beneficial than effort spent sorting and securely disposing of sensitive documents.

**Home Health Paperwork Reduction—Midwest Regional Health System**

Midwest Regional Health System (MRHS) is creating a paperless medical record system for home health practitioners. This will reduce paper usage, create more space, and save time and money. Patients will additionally receive better customer service due to the fact that nurses will spend more time in the field with patients. Although this project will cost $138,000 to implement, the return on investment (ROI) will be less than one month. MRHS estimated a paper savings of 660,000 pieces a year. Combined savings of paper, labor hours, and reduction of mistakes will save the hospital $1.2 million annually. In addition, the nurses will free up capacity to take on new patients for possible added revenue of $864,000/year.

## WASTE HOARDING

Sometimes, stable regulated waste that is difficult to dispose of may be quietly hoarded. This is particularly true of some relatively inert materials whose disposal is regulated as universal waste, such as fluorescent bulbs. Spent fluorescent bulbs can sometimes be set aside for years before their volume becomes large enough that the issue must be dealt with. However, the primary impact of engaging in this kind of hoarding is that the relatively low ongoing expense of dealing with them as they are produced balloons into a single major expense when the hoard is dealt with.

It can be useful to compare the measured amount of hazardous waste with the amount that would be expected based on purchasing and production levels, to see if there is a gap that indicates improper disposal or hoarding. This is probably easiest with a material like fluorescent bulbs that passes unchanged through the system (both entering and exiting as a lightbulb). If fluorescent bulbs are being properly disposed of and purchased, the amount appearing in the waste stream should be *roughly* equal to the amount being purchased.

## TRAINING AND INFORMATION

One of the major barriers to proper regulated waste segregation in hazardous waste, regulated medical waste, and confidential paperwork is often a lack of training. If unsure of the proper disposal method, people will usually err on the side of caution and dispose of the waste in hand using the most secure method available. Also, people usually do not have the option to step away from a task to research the disposal of a waste product of which they are unsure.

Information on waste disposition should be available through as many sources as possible, clear and easily accessible to people with questions. Waste segregation expectations should be clearly defined in the following places:

- As a session in new hire training, where requirements should be laid out as simply as possible
- With signs for waste containers that simply and clearly list what materials should be placed in them
- In the policy manual
- In some kind of readily accessible quick reference guide, possibly involving a flowchart for more complex waste streams
- In employee performance reviews

## ENDNOTES

1. Solid waste is regulated in RCRA Subtitle D. This list of factors is paraphrased from http://www.law.cornell.edu/wex/resource_conservation_and_recovery_act_rcra.
2. RCRA Subtitle C outlines cradle-to-grave responsibilities: http://www.epa.gov/region02/waste/csummary.htm.
3. Cradle-to-cradle certification is available for manufacturers through the Cradle to Cradle Product Innovation Institute: http://www.c2ccertified.org/.
4. The EPA's categorization guidelines of hazardous waste have been paraphrased from http://www.epa.gov/osw/hazard/generation/sqg/handbook/k01005.pdf.
5. For more information on universal wastes, visit the EPA's website at http://www.epa.gov/osw/hazard/wastetypes/universal/.
6. For more information on hazardous waste recycling, visit the EPA's resources: http://www.epa.gov/osw/hazard/recycling/.
7. For more information on hazardous waste recycling, see http://www.epa.gov/osw/wycd/manag-hw/e00-001d.pdf.
8. The EPA has a variety of excellent publications available summarizing many of the subtleties of regulated medical waste: http://www.epa.gov/osw/nonhaz/industrial/medical/publications.htm.
9. This information was provided by St. John Medical Center, included in a training program for medical waste segregation.
10. This information has been paraphrased from the EPA's publication: http://www.epa.gov/osw/nonhaz/industrial/medical/mwpdfs/rx/ch1.pdf.

# 13

## *Afterword: Maybe Don't Call It Green*

There are many terms that can describe waste stream reduction. *Environmentally friendly, sustainable,* and *green* are some of the most common, but these seem to suggest that the environment is something that is external, and that we are somehow acting altruistically when we minimize the environmental impact of our lives. Our resource usage is integrated into that environment, and relies on the sustainability of those natural resources in order to keep functioning. We are in turn reliant on it for our continued survival.

The primary benefit of creating a Lean waste stream is the fact that it shows a positive link between good economic practices and environmental conservation through waste reduction. Beyond garbage, Lean principles can also be applied to a number of other environmental efficiency issues to produce both cost savings and environmental benefits, including transportation reduction, energy use, and water use. This economic benefit is the natural outcome of waste reduction, since waste has costs but no benefits. This relationship is sometimes expressed more fully as the triple bottom line of people, profit, and planet, with meaningful improvements seen as benefitting all three.

The waste stream is often thought of as something external to our processes, in that material only enters the waste stream after it leaves the process. All of the exercises and ideas in this book seek to broaden our perspective on process boundaries, conceptualizing the waste products as still being caught up in process flow. If this perspective is broadened enough, the problem of garbage can be radically reduced and the culture of garbage can be changed to produce a higher quality of life.

# Appendix A: Conducting a Garbage Audit at the University of Oklahoma

The front line of most campus sustainability efforts is its recycling program. Many people who are not otherwise engaged in the field of sustainability may see recycling as the primary ecologically positive activity in their lives, and the program also serves as an important beginning point to educate on the larger principles of resource conservation. But how effective are recycling programs in terms of actual waste recapture?

The effectiveness of recycling can be difficult to assess, since the materials disappear into containers and are processed off site. Because of this, people often either falsely feel that their choice does not matter or assume that they are correctly disposing of something when in fact they are not.

As feedback to the community, university administrations usually provide only high-level statistics on the amount of recyclable materials captured by the system. The University of Oklahoma (OU) recycling office reports overall campus recycling volumes. For fiscal year 2012, the OU recycling department reports processing 1,865,100 pounds of material altogether, noting that this is a 20% increase over recycling volumes in 2011, and that there had been an overall 13% reduction in campus waste between 2011 and 2012.[1] According to OU's environmental health and safety officer, Trent Brown, the school landfilled a total of 6,863,760 pounds of waste during FY 2012, meaning that it successfully recycled a little over 21% of the school's solid waste stream that year.[2] Not only did the waste stream shrink, but a larger portion of it was recycled. The numbers posted for FY 2011 and 2012 also suggest that OU's waste recapture rate increased several percent between the two years as well. In order to make these huge numbers more personal, the OU recycling website also breaks the campus recycling volume down as representing 18 pounds per person (Figure A.1).

These numbers *look* impressive, and there is undeniably a great deal of material being collected here, but these numbers ultimately leave us unable to really assess the health and success of this recycling program or understand the effectiveness of our own personal contributions. The problem is

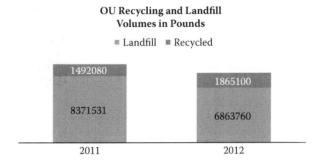

**FIGURE A.1**
Recycling and landfill rates for 2011 and 2012. Note that the overall amount of waste decreases in 2012, and the amount of recycled material increases.

that there is no way to tell how many recyclables remained uncaptured in the 79% of the material that was still landfilled, or how many contaminants were placed in the recycling bins by mistake, only to be landfilled later by the recycling facility. In order to collect performance data that deliver this kind of information, one has to physically go into the waste stream itself to sort and measure its contents alongside the material being recycled.

## CAMPUS GARBAGE AUDIT

For the purposes of this audit, the author selected the National Weather Center (NWC) on the OU Norman campus. The NWC is an excellent location for this audit for several reasons:

- It is a large, physically isolated facility on campus, with a waste stream produced only by its occupants. Its dumpster and recycling collection containers are protected in a loading dock that requires a key swipe to access.
- NWC facilities have collection points for all types of on-campus recycling, including cardboard.
- A broad range of activities take place at the NWC, including classes, research, administrative functions, and the operations of an on-site restaurant.

This audit was conducted purposely without prior notification to the occupants of the NWC, in an effort to capture the by-products of their natural behavior (Figure A.2).

**FIGURE A.2**
The National Weather Center at the University of Oklahoma.

In order to develop a complete picture of the waste streams coming from this facility, it is essential to couple the audit of the garbage with an audit of the material being recycled as well. Auditing the recycling stream generated during the same period examines the effectiveness of NWC's recycling efforts. This comparison not only establishes a true capture/miss rate for recyclables, but also screens for the rate of nonrecyclables that have been placed in the recycling bins improperly.

OU recycling and facilities management were enthusiastic to help with this project, since these data support their mission. As preparation for the audit, every accessible trash and recycling container in the entire NWC facility was emptied. Twenty-four hours following this system flush, these containers were then emptied again and the contents were taken to OU's on-campus recycling facility for sorting. This does not include shred bins for confidential paperwork, which were inaccessible. Because of this, the data on paper production must be skewed low to some degree.

On any given day, 650+ people work in and occupy the NWC, which is approximately a 244,000-square-foot facility.[3] Since the NWC is such a large facility, a day's worth of waste is a significant enough amount to measure for an audit. This is not just because the size of the sample helps get a more accurate waste performance picture per person, but because the size of this facility helps balance out the short waste collection period. One of the most challenging things to measure in this type of audit is material that is used up slowly—such as batteries, lightbulbs, or waste paper accumulating in a shredder—since these are unlikely to appear in any given day's waste. These factors are more normalized in larger populations, where the collection of slowly generated waste more closely approximates actual production.

## MATERIAL ANALYSIS

The first level of material analysis is to simply break down materials by weight following the different available campus recycling options, as seen on the following chart,[4] with an additional category for goods that could have been reused, and a final category for true garbage that could neither be recycled nor reused on campus (Figure A.3).

**FIGURE A.3**
Sorting NWC garbage at the OU recycling facility.

The following table shows a side-by-side comparison by weight of the amount of each type of material disposed of in the garbage vs. in recycling.

| Material Type | Weight Found in Garbage (pounds) | Weight Found in Recycling (pounds) | Percentage Disposed of Properly (%) |
|---|---|---|---|
| Mixed paper | 15 | 31 | 67.4 |
| Cardboard | 26 | 30 | 53.5 |
| Plastics | 10 | 0.9 | 8.3 |
| Aluminum | 3 | 0.5 | 7.0 |
| Steel cans | 5 | 0 | 0 |
| Batteries | 0.4 | 0 | 0 |
| Electronics | 0.9 | 0 | 0 |
| Durable or reusable goods in good condition | 1.6 | 0 | 0 |
| True garbage (neither recyclable nor reusable) | 189 | 0.2 | 99.9 |

This data set is also represented below in graph form (Figure A.4). Occupants of the NWC disposed of approximately 80% of their overall waste correctly, including the true garbage. One positive aspect of these

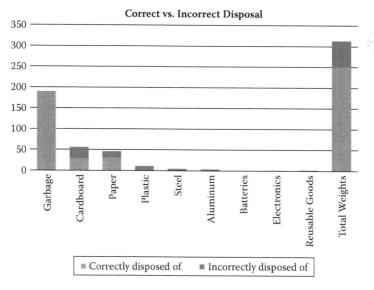

**FIGURE A.4**

Breakdown by category of material in the NWC waste stream, showing which materials were or were not disposed of correctly.

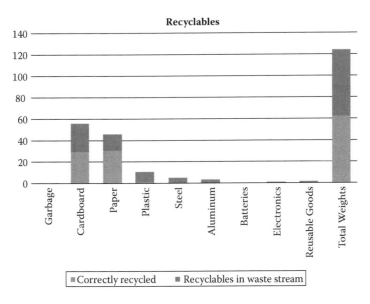

**FIGURE A.5**
Detail of the previous chart, showing the proportions of recyclable materials that were correctly or incorrectly disposed of.

data is the fact that the audit uncovered very little contamination of the recycling stream with garbage. The garbage that did appear in the recycling stream was largely material that people probably legitimately thought could be recycled—such as heavy-grade plastic bubble wrap—and did not represent neglectful contamination from random trash.

However, setting aside the garbage itself, the picture of the recyclable recapture rate is not nearly as good. By weight, a full 50% of the materials that could have been recycled on campus were not, including several material categories (steel cans, batteries, and e-waste) that were not recycled at all (Figure A.5).

In considering these data, it is also useful to consider the presence of water in the solid waste stream. One factor that will distort these data to some degree is the fact that some of the absorbent material in the waste stream was wet, and the water's weight added to its apparent volume. This includes both cardboard and paper in the waste stream, and also the actual garbage itself. Recycled paper and cardboard were probably the only categories that were completely dry. Drink containers recovered from the garbage in this audit also contained an additional 5.5 pounds of liquids, which represents roughly another 2% of the whole sample's weight. Many bags of trash, particularly those containing significant food waste, were pooling with liquid that ran in rivulets away from the sorting area and was not weighed. The presence of this liquid in the garbage increases the risk of spills and injuries while the waste is in the facility, and increases the hauling costs of the waste by adding extra weight and fuel use after it leaves. Therefore, emptying containers before disposal can help reduce the overall cost and energy output of a community waste program.

Since this event only captured material from a single point on campus during a single day, the results of this audit are not necessarily representative of the behavior of the campuswide waste stream as a whole, or even necessarily of this facility over time. The data are accurate but not necessarily extensible. If these data are normal, then the recycling levels would represent 24 pounds of recycled waste per person per day in the NWC, well above the campus norm. In order to develop more robust data, an ongoing monitoring program would need to sample waste from the facility over a longer period. However, these results are generally resonant with the material levels seen in other random samples taken from campus dumpsters over the last several years for use in classroom audit exercises.

---

## EVALUATION

The information on weights and materials leads to an analysis of behaviors, always with the question in mind, "Why was this not recycled?" For the commonly recyclable materials found in this audit, a first-level solution is simply to encourage their disposal in the proper containers through education, transparency, and accountability, or through increased access.

Recycling containers for plastic and aluminum are virtually as common as conventional garbage cans in the NWC, and recycling these materials is no more difficult than placing them in the garbage. A walk-through of the building revealed that virtually every trash can was paired with a plastic/aluminum or paper recycling can (Figure A.6).

On the building's upper floors—housing most of the offices and classrooms—plastic/aluminum cans outnumbered trash cans 2:1, with many halls that had recycling cans for plastic/aluminum and paper, but no regular trash. Occupants generating these recyclable wastes on these

**FIGURE A.6**
Trash can and recycling can in the NWC atrium.

**FIGURE A.7**
Hallway with two recycling cans and no trash can.

floors would have to actively seek out a trash can to avoid recycling it, which is really ideal for a recycling program (Figure A.7).

With all of this accessibility, the low recapture rate must be due primarily to education. According to Heather Murphy, the NWC building administrator, one major barrier is likely linguistic, since all of the signage is in English, and a large number of the students in this facility are

**FIGURE A.8**

The list of unacceptable contaminants from OU recycling attached to this paper recycling container would be daunting to any nonnative English speaker, and is not immediately clear to native speakers. Any kind of barrier like this to quickly understanding what can be recycled will lead to lower participation.

not native English speakers. However, even some native English speakers who were questioned thought that despite labeling, any recyclable material could be placed in any container, to be sorted later by OU recycling, which is not true. A simple, visually clear recycling poster would be of great use in helping people dispose of basic recyclables (paper, plastic, and aluminum) properly, possibly relaying this information in several languages (Figure A.8).

The single highest recyclable material in the waste stream by weight was cardboard. Although there were a number of brown cardboard boxes in the dumpster, the cardboard recovered from the waste stream was primarily not brown shipping boxes, but rather smaller cardboard and chipboard containers (e.g., chipboard food containers, unglazed paper plates, other packaging, etc.). In this case, the problems of education and access were probably relevant in equal measure, since most people may believe that only pure, brown cardboard could be recycled, and may not know where this material should be collected for recycling, since cardboard is routed to a separate stream than either paper or aluminum and plastic and there are no containers set up to collect it within the building. In many facilities on campus, cardboard is simply stacked on the floor next to the recycling bin, or at some other informal location; small cardboard is effectively not collected at all. Because of this, virtually all of it is mixed with the trash

**FIGURE A.9**
The NWC's cardboard recycling trailer and dumpster. These trailers are parked at locations around campus and are collected whenever they are full. A map of all of the trailer locations is available on the OU sustainability website.[5]

and ends up in the dumpster, despite the fact that the dumpster sits next to the cardboard collection trailer (Figure A.9).

Beyond these regular recyclable materials, this audit also uncovered a few other items in the trash that could be recycled or reused. Batteries and electronics are two material categories that OU recycling collects, although most departments do not have a system set up to collect and hand them off. In this audit, the reusable materials that were uncovered were really consumables that had not been fully consumed: ballpoint pens, a roll of unused garbage bags, a mostly full can of spray paint, etc. More often, reusable materials in the waste stream are durable goods that are still capable of serving their original function, such as three-ring binders. These items are often seen as being easier to throw away than store or find a new home for.

The examination of paperwork found in the garbage very often turns up sensitive personal information that could be used for the purposes of identity theft. Many people have a sense of false security around garbage, incorrectly feeling that material placed in the trash is disposed of without the likelihood of any future human contact. This audit uncovered very little sensitive material in the waste stream, with only one document producing a name, address, and insurance policy number. Paperwork that had been placed in the recycling stream was shredded and baled securely by OU recycling.

Many of the behaviors observed through the garbage analysis were confirmed by interviewing graduate students from the meteorology department. Although there are large recycling containers in the public spaces that are emptied by the custodial staff, this staff does not have the manpower to empty the containers in each individual office, where most waste is generated: "This makes people less likely to use them, since they have to carry the full bins out of their office to a larger recycling bin somewhere in the hallways, where they might have to throw the cans and bottles in one by one." Apart from this, there was not always a clear understanding of what could be recycled, especially with paper and cardboard products: "I wish I could easily recycle the cardboard cups and cardboard heat protectors that coffee is served in. I and many other students drink a lot of coffee, and I can imagine that the volume of coffee cups is huge. It's possible they can be recycled in one of the current bins, but that isn't made clear."[6]

## CONCLUSIONS

One of the difficulties with any recycling program is enforcement. Although recycling is widely available across campus, there is no accountability for those who choose not to use it, since the contents of the garbage itself are not monitored in any way. When this is coupled with even slightly unclear material recycling policies, the participation rates for all but the most basic materials will be low except for people who feel passionately enough about sustainability to take the extra step of seeking out information about the program.

Circulating the results of this audit among NWC staff and students might result in a spike in all kinds of recycling participation, since it quantifies and provides feedback on their performance, demonstrating that the waste stream is no longer invisible. However, that spike would be transient unless it was continually reinforced with further feedback, and it might actually have a negative effect, merely fostering paranoia and resentment toward the "garbage police." Although the information gathered here is interesting, it is best used to drive educational policy rather than to expose undesirable behavior.

For this analysis, data were collected from the University of Oklahoma out of sheer convenience. OU is known as a leader in sustainable practices in the state, and the author had easy access to its waste stream and internal data. Whether results similar to these might be found in assessing the waste stream of any major educational institution is an open question, and the general lack of knowledge of waste stream contents makes it difficult to project how close these results are to the norm.

## ENDNOTES

1. See http://www.ou.edu/content/sustainability/currentpractices/recycling.html.
2. Information obtained by the author in a personal email, October 31, 2013.
3. Information on the NWC can be found on the OU School of Meteorology website: http://som.ou.edu/.
4. The following website details what can be recycled on campus: http://www.ou.edu/sustainability/currentpractices/recycling.html. Other categories of materials can also be recycled on campus, including pallets and toner cartridges, but no examples of these appeared in the material audited.
5. See https://alumni.ou.edu/content/facilities/sustainable_practices/recycling/cardboard_trailer_locations.html.
6. Quotes obtained by the author in a personal email, December 5, 2013.

# Appendix B: Norman, Oklahoma, Municipal Compost Facility

The City of Norman, Oklahoma, provides an excellent example of a municipal compost operation. Norman has operated a compost facility since the mid-1990s, and in that time it has become a city service that Norman residents highly value. Norman's facility is the most established municipal compost yard in the state, and representatives from other cities often tour it. Finished compost produced by the facility is given back to citizens at no cost approximately every 100 days.

In Norman, three different kinds of household waste are collected weekly at curbside: garbage, single-stream recyclables, and green waste. The green waste collection schedule changes with the seasons, scaling back to a single monthly collection during the winter months when the production of yard waste is lowest. Garbage and recycling are collected on the same weekday, with recycling collection contracted out to a private company. Both of these types of waste are collected using 95-gallon containers that are picked up by a truck with a hydraulic arm. Green waste is picked up manually on a different day of the week, and the city does not issue standard containers for it.[1] In addition to curbside pickup, yard waste can be brought directly to the facility by either citizens or contractors and dropped off at no cost.

Although this collection schedule requires two separate sweeps of the city by the sanitation fleet, city officials do not feel that this represents a significant increase in transportation. This is because the city would be collecting the same amount of waste material with or without the compost facility, so if yard waste were collected as part of household waste, the extra trucks and labor currently dedicated to yard waste pickup would be required for the conventional waste instead.

The City of Norman spells out green waste collection in detail on its website, including exact specifications for what kinds of waste are accepted and how that waste is to be prepared for collection. In its literature, this waste is referred to specifically as "yard waste" rather than "green waste" as an indication that it does not currently accept any green kitchen waste through this system. Theoretically, any kitchen waste free from animal fats could be included in this mix and still be processed properly, but the risk of contamination is very high. The city sends out reminders to residents following holidays that special organic wastes—such as jack-o-lanterns or Christmas trees—are acceptable for composting.

During the 2012 fiscal year, Norman's compost facility collected 8,611 tons of waste, producing a direct savings of $162,061 from avoided landfill tipping fees. This cost savings represents a little over a 10% reduction in the city's overall waste disposal costs based on the tonnage of waste land-filled. The composted material was split fairly evenly between that which was collected curbside and that which was brought in and dropped off. Although compost is given away at no cost from this facility, the city does offer to load it into trucks at a cost of $10 per backhoe scoop. The revenue from this loading provided an additional $25,390 in 2012. While this revenue does not cover the costs of collecting and processing the waste, it should be remembered that the bulk of the processing costs would have been incurred anyway if the yard waste had been sent to the landfill.

Compost requires a source of water in order to cook, and as part of its comprehensive environmental plan, the City of Norman is pursuing permitting that would allow it to water the compost using sterilized effluent from the municipal wastewater treatment plant. Regulation of this water use is governed by the Department of Environmental Quality, which oversees this type of facility in terms of its potential ability to impact groundwater quality (Figure B.1).

Municipal compost yards can be established following Norman's model with minimal capital costs. For its first few years of operation, Norman's compost yard was primarily maintained by borrowing equipment from other departments, minimizing the facility start-up costs. Economies of scale will apply to this process, but City of Norman officials estimate that cities with populations as low as 10% of Norman's (approximately 10,000) could cost-effectively set up a compost yard following this model.

**FIGURE B.1**
Compost in process at the City of Norman facility. Some of the equipment required for turning and managing the compost is visible in the background.

## ENDNOTE

1. See http://www.ci.norman.ok.us/utilities/sa/sanitation-yardwaste-collection.

# Appendix C: Regulated Medical Waste at St. John Medical Center

Probably the best way to see regulated medical waste regulations in action is to look at the waste practices of an exemplary medical facility. St. John Medical Center (SJMC) in Tulsa, Oklahoma, carries out waste disposal at a level that matches the best practices for the field, following the Environmental Protection Agency's (EPA) national guidelines. There are no additional state-imposed regulatory guidelines in Oklahoma, so SJMC's performance would track to the baseline regulations for medical facilities in any state.

SJMC conducts annual refresher training for its staff in the management of regulated medical waste, and has clear signage posted in its facility as to exactly what may be placed in any type of container. Although each container is labeled individually, the posted list shown in Figure C.1 appears frequently, covering every major category of waste.

In practice, SJMC breaks down its waste into four major categories, and then further subdivides some of those categories. All waste types are color-coded, making it easy to tell the material that goes in any container regardless of the container's shape or size:

- **Regulated medical waste (red bags):** Regulated medical waste is processed throughout the facility in red containers. This material is collected using red biohazard garbage bags in red containers, which flow together into 96-gallon roll-off containers that are then collected daily and autoclaved to sterilize the contents. Red sharp containers enter this stream as well (Figures C.2 and C.3).

    Not all materials that have come in contact with human blood are automatically considered regulated medical waste. The primary consideration is the ability of the blood to leak out and contaminate other materials. For example, a bandage with a spot of dry or absorbed human blood can be placed in the conventional waste stream, but a

**Pharmaceutical Waste Segregation Chart: Nursing**

| Return to Pharmacy Waste | RCRA & Pharmaceutical Waste | Inhalers | Drain | Regulated Medical Waste | Sharps |
|---|---|---|---|---|---|
| Partial narcotics on this list must be double witness wasted and placed in pyxis return to pharmacy bin: | **No Controlled Substances** | **Place in plastic bag and send through tube to return to pulmonary services.** | **All Controlled Substances** *(except RCRA hazardous: diazepam, fentanyl patches, paregoric, chloral hydrate, and epidural cassettes if unable to drain)* | Contaminated or dripping with blood/blood products | • Needles<br>• Scalpel blades & lancets<br>• Glass slides and tubes<br>• Empty glass medication vials and ampoules<br>• Electrocautery tips<br>• Disposable suture sets and biopsy forceps |
| Injectable diazepam | Place all partial non-narcotic meds in this container, including: | Includes all compressed gas inhalers (if not sent home with patient) | | Saturated or grossly soiled disposables, i.e., bloody gauze, dressings, lap pads, surgical peri-pads, gloves | |
| Fentanyl patches (used only) | Partial **medicated** IV bags | | Normal Saline (NS, ½ NS) | Containers or tubes with fluid blood or blood products not | **Partial Biohazardous Pharmaceuticals:** |
| Paregoric | Pills/Liquid PO Used patches (**except fentanyl**) | | Dextrose | discarded or flushed i.e., blood sets, suction canisters, drainage sets | Abraxane    Albumin<br>Aranesp    Botox |
| Chloral Hydrate | **Note:** All P-listed waste meds **and packaging** | | Lactated Ringers | | Botox    Digifab |
| Epidural cassette (if cannot be drained) | must be placed in a clear plastic baggie prior to disposal in this container, including: | | D5W, D10W, D50W | Lab specimens, non-breakable tubes | Flumist    Gammagard<br>Hepagam B M-M-R II |
| | | | IV fluids with electrolytes | | Privigen    Recothrom |
| | •Coumadin (Warfarin) | | Electrolytes | Blood spill clean-up materials | Thrombin-JMI |
| | •Nicotine | | **No Other Meds or IV Fluids Can Be Flushed** | | Thymoglobulin |
| | •Physostigmine | | | No liquids | Varivax    Winrho SDF<br>Zostavax |

June 2013                    Questions? Please Contact the Pharmacy Department.

**FIGURE C.1**
St. John's medical waste segregation chart.

**FIGURE C.2**
Container for regulated medical waste.

grossly soiled bandage that could drip blood—or would drip blood if squeezed—must be treated as regulated waste. In Figure C.4, the containers in this conventional trash can contain blood products, but a coagulant has been added rendering the blood solid and unable to spill.

**FIGURE C.3**
Red bags and small containers placed in a red biohazard bin.

**FIGURE C.4**
Chemically coagulated blood in the conventional waste stream.

- **Chemotherapy (yellow containers):** This stream contains any items that were used in the administering of chemotherapy, and may have been trace contaminated with chemotherapy drugs (Figure C.5).
- **Conventional waste (clear bags):** Conventional garbage that is neither chemically hazardous nor a biohazard is sent through a trash compactor, and then to the landfill. These bags should not contain any regulated waste, but may contain items that have come in contact with human waste (such as urine cups or stool containers). These bags are clear because opaque bags had facilitated employee theft of medical supplies in the past (Figure C.6).

**FIGURE C.5**
Chemotherapy waste collection point.

**FIGURE C.6**
Conventional waste in transparent bags.

**FIGURE C.7**
Blue bags of linens on carts headed to washing facility.

- **Soiled linens (blue bags):** Soiled linens, collected and processed on site. The carts that are used to collect soiled linens are also washed and sanitized before they are allowed to reenter the facility (Figure C.7).
- **Hazardous waste:** Hazardous chemical waste produced by medical facilities runs the gamut from paint thinners and radioactive material to batteries and waste medicines. Black boxes are used to collect all nonnarcotic pharmaceuticals that are not administered to patients. The disposal of any waste narcotics is strictly regulated by the Drug Enforcement Administration (DEA) (Figure C.8).

This overview is a brief look at the actual practice of implementing Resource Conservation and Recovery Act (RCRA) standards in a comprehensive medical facility, but it highlights the complexity of dealing with this waste stream. Many of the principles seen in this case would also apply equally well to nonmedical segregated waste disposal.

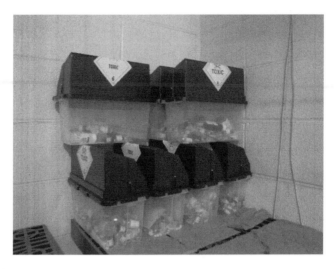

**FIGURE C.8**
Nonnarcotic pharmaceutical waste collection.

# Index